MATHEMATICS
FOR PRACTICAL USE

*the text of this book is printed
on 100% recycled paper*

ABOUT THE AUTHOR

Kaj L. Nielsen received the degrees of B.S. from the University of Michigan, M.A. from Syracuse University, and Ph.D. from the University of Illinois. He has held teaching positions at Syracuse, Illinois, Brown, Butler, Purdue, and Louisiana State University. He has been associated with a number of industries as a research engineer and scientist. Presently, he is Director of the Systems Analysis Division at Battelle Memorial Institute in Columbus, Ohio. Dr. Nielsen has written numerous articles based upon original research in mathematics and published in leading mathematical and engineering journals. He is the author or coauthor of *College Mathematics, Plane and Spherical Trigonometry, Problems in Plane Geometry, Logarithmic and Trigonometric Tables,* and *Differential Equations* in the College Outline Series, and of three other mathematical books.

EVERYDAY HANDBOOKS

MATHEMATICS
FOR PRACTICAL USE

Kaj L. Nielsen

BARNES & NOBLE BOOKS

A DIVISION OF HARPER & ROW, PUBLISHERS

New York, Hagerstown, San Francisco, London

77 78 79 80   12 11 10 9 8 7

Manufactured in the United States of America

TO
GLADYS

PREFACE

As we advance in our state of civilization we are continuously reminded of the need for a knowledge of mathematics. The calculation of discounts in sales, amounts of commissions, measuring for drapes, reading a gas meter, calculating the miles per gallon, baseball averages, and interest rates are only a few examples of the everyday needs.

It is not particularly difficult to acquire a reasonable skill in the use of elementary mathematics. This book has been written to help you acquire that skill and at the same time point out the numerous applications to typical problems which occur around the house, in the operation of the automobile, and in general trying to live comfortably in our present society. To this end the book has been written especially for those who have only a general knowledge of arithmetic and for those who have forgotten even the simplest concepts or have never had the opportunity to master the fundamentals of mathematics.

The book begins with the basic concept of numbers and then treats progressively more complicated mathematical operations from simple addition to the solving of equations, the concepts of geometry, the use of graphs, interest and annuity computations, the application of logarithms, and the use of tables. The problems and illustrations are chosen from everyday experiences of people in all walks of life. The emphasis is on methods for solving the problems; these methods are simple and nontechnical but thorough and authoritative.

It is hoped that the reader will discover that mathematics is not a dull subject, consisting merely of laws and rules, but that it is very much alive and a part of all of us. Mathematics has long been recognized as the foundation of all the sciences; it is also the foundation of business, government, and the very existence of human beings. In the later chapters of the book we discuss the problems of running a home and the mathematical problems of life in general.

To test your skills we have placed a number of problems throughout the book, and the answers are provided in the back. In order to benefit from the learning it is necessary to exert the effort to solve these problems, and by developing the skill one may also find a pleasure and become interested in pursuing the subject further.   An appendix has been added to stimulate interest and to give an introduction to the entertainment which can be derived from a knowledge of mathematics.

Tables and statistical facts which have been included also make this book a good source material for quick reference.

K. L. N.

TABLE OF CONTENTS

LIST OF HANDY ANDY'S

Chapter I

ARITHMETIC

Introduction

Mathematics is, in a sense, a language. It is based on definitions of words, symbols, and signs. It is a universal language that is the same throughout most of the world. Whenever possible it gives the same meanings to the words as those used in our everyday spoken language. There will be different connotations, and these must be studied and learned by the reader. Should the reader fail to memorize the definitions and truly make the language of mathematics a part of his everyday language, he will have difficulty in understanding the subject and solving the technical problems.

Mathematics is an exact science based upon many years of development. As each principle is established it remains a part of the language. Mathematics is still growing as each year more and more is being "created." In fact, it has been said that more mathematics has been developed in the last 50 years than in all the preceding years. Mathematics is based on logical procedures. Thus we move from one step to another with clear thinking and well-defined logical steps.

Over and above the fact that mathematics is tremendously important in our modern civilization, it is also interesting. A certain amount of satisfaction should be obtained from arriving at a correct answer. Although this book is devoted to the teaching of mathematics which is necessary to solve practical problems, it is hoped that the reader will also enjoy the subject matter and a certain group will be inspired to pursue it further.

The Number System

We shall begin our study of mathematics with the number system. Although there are two common systems used to express numbers, the Arabic and Roman, the Arabic is the more prominent. Each consists of a set of figures known as the alphabet of the notation.

2

ARABIC		ROMAN	
0	Zero	I	One
1	One	V	Five
2	Two	X	Ten
3	Three	L	Fifty
4	Four	C	One Hundred
5	Five	D	Five Hundred
6	Six	M	One Thousand
7	Seven		
8	Eight		
9	Nine		

In the Arabic notation the ten figures are called **digits**. By combining these digits we can obtain all the numbers needed for our application to practical problems. In order to speak of the numbers we give names to the various parts of the number. If a number has *one* digit it is of the first order and is read as **Units**. If it has *two* digits it is of the second order and is read as **Tens**. The names are illustrated by analyzing the following whole number.

$$4, 376, 982, 134, 653$$

ORDER	NAME	DIGIT
1st	units	3
2nd	tens	5
3rd	hundreds	6
4th	thousands	4
5th	ten thousands	3
6th	hundred thousands	1
7th	millions	2
8th	ten millions	8
9th	hundred millions	9
10th	billions	6
11th	ten billions	7
12th	hundred billions	3
13th	trillions	4

The number is read: *four trillion, three hundred seventy-six billion, nine hundred eighty-two million, one hundred thirty-four thousand, six hundred fifty-three.*

If a decimal point is placed in a grouping of digits, we are indicating parts of a unit. Again names are given to these parts. Thus:

28, 432. 028

Fractional Part of a Single Unit		thousandths	8
		hundredths	2
		tenths	0
		decimal point	.
Whole Number		units	2
		tens	3
		hundreds	4
		thousands	8
		ten thousands	2

To express a number in the Roman notation we first select the letter of highest value contained in the number, place to the right of it the letter of the next largest sum, and continue this process until we have all the numbers. Let us consider for example the number 137. The Roman notation for it is obtained as follows:

$$
\begin{array}{rccccccc}
100 = & C \\
10 = & & X \\
10 = & & & X \\
10 = & & & & X \\
5 = & & & & & V \\
2 = & & & & & & II \\
\hline
137 = & C & X & X & X & V & II
\end{array}
$$

In order to save time and space, numbers like 4, 9, 40, 90, etc. are contracted by using two letters whose difference is equal to the number.

$$
\begin{array}{rll}
4 = & IIII & = IV \\
9 = & VIIII & = IX \\
40 = & XXXX & = XL \\
90 = & LXXXX & = XC \\
400 = & CCCC & = CD \\
900 = & DCCCC & = CM.
\end{array}
$$

There are a number of definitions of terms associated with numbers which are already familiar to the reader. Some of them are now listed in order that they may be easily recalled to mind.

If a number does not specify anything definite, it is called an **abstract number**. *Examples:* 2, 7, 19, 23.

If a number is connected with a definite object, it is called a **concrete number**. *Examples:* 3 inches, 60 miles, 7 boys.

An **integer** is a whole number. *Examples:* 3, 43, 1776, 148933.

A **factor** of a whole number is a whole number that divides it exactly. *Examples:* 3 is a factor of 15; 4 is a factor of 20.

An **even number** is an integer that is exactly divisible by 2. *Examples:* 2, 4, 6, 8, 42, 1028.

An **odd number** is an integer that is not exactly divisible by 2. *Examples:* 1, 3, 5, 7, 9, 11, 33, 57.

A **prime number** is a number which has no factors except 1 and itself. *Examples:* 1, 2, 7, 11, 13, 23.

A **composite number** is a number that is not prime. *Examples:* 8, 33, 45, 1028.

A **common divisor** of two or more numbers is a factor that will exactly divide each of them. *Example:* 3 is a common divisor of 9 and 27.

If a common divisor is the largest factor possible, it is called the **greatest common divisor.** *Example:* 9 is the g.c.d. of 9 and 27.

If a number is exactly divisible by two or more numbers, it is a **common multiple** of them. *Example:* 18 and 36 are common multiples of 2, 3, and 9.

If we wish to express a part of a unit, we call upon a quantity called a **fraction.** There are two ways of expressing a fraction:

$$common\ fraction:\ \tfrac{1}{2},\ \tfrac{2}{3},\ \tfrac{7}{6};$$
$$decimal\ fraction:\ .5,\ .375,\ 1.236.$$

Each manner of expression is an indicated division. In the common fraction the number above the line is called the **numerator** and the number below the line is called the **denominator.** In the decimal fraction the denominator is understood to be ten ($.5 = \tfrac{5}{10}$), hundred ($.25 = \tfrac{25}{100}$), thousand ($.125 = \tfrac{125}{1000}$), etc.

A **proper fraction** is one in which the numerator is smaller than the denominator; e.g., $\tfrac{2}{3}$.

An **improper fraction** is one in which the numerator is bigger than the denominator; e.g., $\tfrac{7}{6}$.

A **mixed number** is one which contains a whole number and a fraction; examples: $2\tfrac{1}{2}$, $7\tfrac{2}{5}$, $1\tfrac{3}{4}$.

The number system can be expressed geometrically as distances from a fixed point on a straight line, provided a convenient unit of measure is chosen.

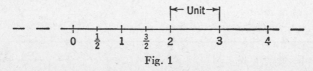

Fig. 1

The question may then be raised, "What does a distance to the left of the point 0 represent?" It is a valid question, and the answer is that it represents a **negative number** which is denoted by a minus sign; thus "−2." The reader is familiar with negative numbers through the reading of a thermometer, where 10 degrees below zero is usually denoted by −10°. To distinguish between a negative and a positive number we use the minus sign, "−," for *negative* numbers and the plus sign, "+," or *no sign* for the *positive* numbers.

The **absolute** or **numerical** value of a number is the value that the number has without regard to its sign and is usually denoted by two vertical bars | 5 |. For example, we have

$$\lceil -5 \rceil = 5 \quad \text{and} \quad | 5 | = 5.$$

This is by no means the complete number system but it forms a good basis on which to begin. Some of the other parts will be explained later in the book.

HANDY ANDY No. 1	
A number is divisible by	if
2	last figure is 0 or one divisible by 2.
3	sum of its digits is divisible by 3.
4	number represented by last two digits is divisible by 4 or both are zeros.
5	last digit is 0 or 5.
6	it is an even number the sum of whose digits is divisible by 3.
7	(no known rule)
8	number represented by last three digits is divisible by 8.
9	sum of its digits is divisible by 9.

Symbols and Signs

A part of mathematics is a shorthand of symbols and signs which we must learn as a part of the language. The basic ones are:

I. *The signs of operations:*
    (a) Addition: +; e.g., $3 + 2$
    (b) Subtraction: −; e.g., $5 − 3$
    (c) Multiplication: × or ·; e.g., $7 × 3$ or $7 · 3$
    (d) Division: ÷; e.g., $4 ÷ 3$

II. *The signs of order:*
  (a) Is less than: $<$; e.g., $4 < 9$.
  (b) Is greater than: $>$; e.g., $7 > 5$.
  (c) Is equal to: $=$; e.g., $7 = (14 \div 2)$.
III. *The signs of grouping:*
  (a) The parenthesis: ( ); e.g., $(3 + 2) + (4 + 7)$.
  (b) The brackets: [ ]; e.g., $[(3 + 2) + 5] - 6$.
  (c) The braces: { }; e.g., $\{[8 + (3 - 1)] - (6 + 2)\}$.

In the common fraction, the line denotes a division; for example, $\frac{2}{3}$ means 2 divided by 3. The parenthesis is also used to denote a multiplication; for example, $(2 + 3)(7 - 4)$ means $(2 + 3) \times (7 - 4)$.

Addition

The operation of adding two numbers is a process of regrouping. Thus to add 7 and 8 we regroup the numbers to obtain 15. We have not increased the number of objects but have simply found another way of expressing 7 and 8. The terms of addition are:

$$\text{Addend} + \text{Addend} = \text{Sum (Total)}.$$

In order to perform more complex addition we must first be able to combine any two of the digits in any order; i.e., we must know $3 + 4 = 7, 5 + 8 = 13, 9 + 6 = 15$, etc. Write down the integers 1, 2, 3, 4, 5, 6, 7, 8, 9, 0 across a page; then add 0 to each of these; 1 to each; 2 to each; etc. There are 100 such combinations. Further skill is developed by knowing the combination of a *two-digit* number with all the digits. Thus we should be able to combine 27 and 9 to obtain 36; $17 + 8 = 25$; $87 + 9 = 96$; etc.

To add groups of numbers we first place them in vertical columns such that the units, tens, hundreds, etc. are directly under digits of similar order. We then add each column starting at the right and "carrying" a figure in the sum from one column to the next.

Examples

(a)	(b)
1232	
32475	1898
2394	32
86	461
952	3009
87318	27
Sum = 123225.	Sum = 5427.

In example (a) the sum of the right column is 25; we write the 5 at the bottom of the same column and place the 2 at the top of the next column to the left. In adding the numbers of this next column we then include the 2 in the sum to get 32, and place the 2 in the tens column and carry the 3 to the hundreds column. The process is continued until all numbers are used. In example (b) the number which was "carried" is not written down but is immediately added mentally to the first number in the next column, thus saving time.

To add numbers containing decimals make certain that all the decimal points are placed in the same column.

Examples

(a)	(b)
3.054	$ 125.57
17.3	1632.63
3.47	5.75
23.824	$1763.95

If it is necessary to add a rather long column of figures, it may be advisable to split it into two or more parts and then add the partial sums.

Examples

(a)		(b)	
2527		$1639.41	
192		325.67	
4638		17.32	
7215		439.16	$2421.56
1385		13.92	
2367		7.85	
1389		1600.27	
1437	21,150	391.32	2013.36
1663		1.49	
3972		6.72	
4867		7.39	
2394		42.63	
2936		19.87	78.10
4275		132.63	
2739		349.31	
293		431.69	
1971	25,110	127.33	1040.96
	45,260		$5553.98

A long addition should always be checked, and this is best accomplished by adding the numbers in reverse order; that is, if the numbers were first added by starting at the top and adding downward, then to check start at the bottom and add upwards. Although it is usually difficult for most people to add numbers horizontally, it is, nevertheless, advantageous to be able to do so and some proficiency can be attained by considerable practice.

Problems I.

1. (a)	(b)	(c)	(d)	(e)
32	398	8148	$15.32	$1649.33
98	672	6732	7.69	139.67
67	349	9355	3.42	455.79
33	676	8378	9.61	322.13
55	897	1377	11.46	157.63
67	341	4322	3.55	17.28
78	692	6938	2.35	329.42

2. Add both horizontally and vertically.

Items	(a)	(b)	(c)	(d)	(e)	Totals
A	$1.67	$3.22	$0.67	$1.93	$2.55	
B	0.39	1.63	1.23	2.67	1.45	
C	3.29	1.39	0.13	1.63	3.19	
D	0.15	2.55	0.67	1.87	4.27	
E	0.75	1.45	0.39	1.69	0.63	
F	1.32	0.19	0.81	2.18	1.38	
Totals						

3. If there are two possible highway routes between two cities and the map shows the following mileage along each route:

Route A: 23, 7, 5, 2, 13, 19, 46, 3, 2, 6;
Route B: 34, 9, 17, 21, 3, 6, 19, 12, 1, 8;

which is the shorter route?

4. To put an article on the market the labor costs $27.50, the material costs $9.75, the advertising costs $1.50, and the transportation costs $2.65. What should be the price of the article if a profit of $17.75 is desired?

5. On a certain hole a golfer hits a ball 233.75 yards, 194.9 yards, 62.65 yards, 18.5 yards, and 1.2 yards. How far did the golf ball travel?

6. In the 1960 presidential election 10 states had electoral votes of 45, 32, 32, 27, 25, 24, 20, 16, 16, and 14. How many of the 537 votes did they control?

Subtraction

The process of "taking something away" is called subtraction. In a broader sense there are three kinds of questions which lead to the process of subtraction:

(a) *How much is left?* Mary spends 35¢ of her 50¢; how much has she left?

(b) *How much is the difference?* One car costs $3375.50 and another costs $2942.75; how much is the difference in price?

(c) *How much more is needed?* Tom has $23.50 and would like to buy a bicycle for $36.75; how much more does he need?

Subtraction is the "opposite" process from addition.

Minuend − Subtrahend = Remainder.

The first step in the study of subtraction is to be able to recognize the answer instantly and automatically when subtracting a one.digit number from a one digit number. Arrange the digits 0, 1, 2, 3, 4, 5, 6, 7, 8, 9 across the page and subtract from each first 0, then 1, then 2, etc. If we allow only positive answers, there are 55 such combinations.

To subtract one large number from another we again arrange the two numbers such that the units are in the same column, the tens are in the same column, etc.

Examples

(a)	(b)
1897	396.23
543	85.02
1354	311.21

Should the number below in any column be greater than the number above it, then 10 units are added to it and the subtraction is resumed. This operation reduces the number to the left by 1 and is called "borrowing."

Examples

(a)	(b)

0 12 13 17   8 12 11   11 12

~~1~~~~3~~~~4~~7   1~~9~~~~3~~~~2~~.~~2~~2

     8 6 9   8 4 6 . 3 5
-----------   --------------
     4 7 8   1 0 8 5 . 8 7

When we subtract by taking away and borrowing, we should learn to do so without writing the changed numbers:

	7 from $14 = 7$
6324	8 from $11 = 3$
987	9 from $12 = 3$
5337	0 from $5 = 5$

The simplest method of checking subtraction is to add the remainder to the subtrahend and the sum should equal the minuend.

Example

	$8396.35	minuend
subtract:	1657.68	subtrahend
	$6738.67	remainder
add:	1657.68	subtrahend
	$8396.35	minuend

It should not be necessary to rewrite the subtrahend a second time as was done in the example, but rather the process of addition should be done mentally.

If the subtrahend contains more digits after the decimal point than the minuend, we annex the necessary number of zeros to the minuend and borrow as before.

Example

Find $3.63 - 1.847$.

Solution.   Write

          3.630
          1.847

          1.783

Problems II.

1. 1575.326	2. $3275.67	3. 4632.27
$-$ 983.154	$-$ 2986.75	$-$ 3176.27

4. $96.43 - 48.52$.   5. $367 - 142$.   6. $19.38 - 7.632$.

7. If you purchase articles costing 25¢, 39¢, $1.78, 69¢, 87¢, $2.94, and 42¢, how much change should you get from a ten dollar bill?

8. The annual report of an airline stated that they carried 875,632 passengers last year compared with 639,747 the year before. What was the increase in passengers carried?

9. If withdrawals of $47.63, $9.19, $147.68, $397.34, and $118.52 were made from an original bank account of $2000.00, what was the balance after each withdrawal?

10. From a board 16 feet long, a carpenter cuts 4 boards of length 3, 2, 5, and 4 feet. If each cut wastes 0.0052 feet (approximately $\frac{1}{16}$ inch), how long is the piece left over?

Multiplication

The process of multiplication is a short cut for addition of numbers of the same kind. Thus to add 72 six times, we shorten the addition by multiplication.

Multiplication	Addition
72	72
6	72
$\overline{12}$	72
42	72
$\overline{432}$	72
	72
	$\overline{432}$

The terms of multiplication are:

Multiplicand × Multiplier = Product.

Before we can be proficient in multiplication it is necessary to *memorize* the product of each digit by every other digit. These simple multiplications can be verified by additions.

Examples

(a) $3 \times 2 = 6$ since $3 + 3 = 6.$

(b) $8 \times 3 = 24$ since $8 + 8 + 8 = 24.$

(c) $7 \times 5 = 35$ since $7 + 7 + 7 + 7 + 7 = 35.$

The process of memorization can be aided by the use of a multiplication table which we shall extend up to the number 12.

MULTIPLICATION TABLE

	0	1	2	3	4	5	6	7	8	9	10	11	12
0	0	0	0	0	0	0	0	0	0	0	0	0	0
1	0	1	2	3	4	5	6	7	8	9	10	11	12
2	0	2	4	6	8	10	12	14	16	18	20	22	24
3	0	3	6	9	12	15	18	21	24	27	30	33	36
4	0	4	8	12	16	20	24	28	32	36	40	44	48
5	0	5	10	15	20	25	30	35	40	45	50	55	60
6	0	6	12	18	24	30	36	42	48	54	60	66	72
7	0	7	14	21	28	35	42	49	56	63	70	77	84
8	0	8	16	24	32	40	48	56	64	72	80	88	96
9	0	9	18	27	36	45	54	63	72	81	90	99	108
10	0	10	20	30	40	50	60	70	80	90	100	110	120
11	0	11	22	33	44	55	66	77	88	99	110	121	132
12	0	12	24	36	48	60	72	84	96	108	120	132	144

This table is read in the following manner:

$$8 \times 4 = 32.$$

Note that

$$4 \times 8 = 32.$$

In other words, the order of multiplication is immaterial.

$$8 \times 4 = 4 \times 8 = 32.$$

To multiply two large numbers we choose the smaller number as the multiplier and arrange the work as illustrated in the following example:

```
2463  Multiplicand
 276  Multiplier
14778 ⎫
17241 ⎬ Partial Products
4926  ⎭
679788 Product
```

The multiplier is placed below the multiplicand so that units, tens, etc. of the numbers are in columns. The process is started by taking the units digit of the multiplier (6) and multiplying it with all the digits of the multiplicand from right to left. Thus $6 \times 3 = 18$ and we place the 8 of this product below the 6 of the multiplier and add the 1 to the next product of 6 times 6. Thus $(6 \times 6) + 1 = 36 + 1 = 37$. Place the 7 of this operation to the left of the 6 and add the 3 to the next product of 6 times 4; thus $(6 \times 4) + 3 = 27$. Place the 7 of this operation to the left of the previous 7 and add the 2 to the product of 6 times 2; thus $(6 \times 2) + 2 = 14$. Place the 14 to the left of the 7 and this completes the multiplication with the units digit of the multiplier. Repeat the same procedure with the tens digit of the multiplier and all the digits of the multiplicand. The first product of the second line is placed in the column with the tens digit of the multiplier. When the second line of products is completed, proceed with the third line of products using the hundreds digit of the multiplier and all the digits of the multiplicand. The first product of the third line is placed in the column with the hundreds digit of the multiplier. This process is continued until all the digits of the multiplier have been used. These products are called *partial products*, which are now added to obtain the final product.

The multiplication of numbers containing decimals is the same as explained above except that we have to determine the place of the decimal point in the product. To find this position *add* together the number of decimal places in the multiplicand and the multiplier; this is the number which must be "pointed off" in the product, starting from the extreme right and counting to the left. Should the number of places needed for the decimal point be larger than the number of digits in the product, prefix as many zeros as needed.

Examples

(a)	(b)
2 2.7 5	.2 5
3.1 5 1	.0 4 1
‾‾‾‾‾‾‾	‾‾‾‾‾‾
2 2 7 5	2 5
1 1 3 7 5	1 0 0
2 2 7 5	‾‾‾‾‾‾‾‾
6 8 2 5	.0 1 0 2 5
‾‾‾‾‾‾‾‾‾‾‾	
7 1.6 8 5 2 5	

A multiplication is checked by interchanging the multiplicand and the multiplier.

Example

Multiply: 3 2.1 2   Check: 7 6.8
                7 6.8   3 2.1 2
              ‾‾‾‾‾‾‾   ‾‾‾‾‾‾‾
              2 5 6 9 6   1 5 3 6
              1 9 2 7 2   7 6 8
              2 2 4 8 4   1 5 3 6
              ‾‾‾‾‾‾‾‾‾   2 3 0 4
              2 4 6 6.8 1 6   ‾‾‾‾‾‾‾‾‾
                                               2 4 6 6.8 1 6

HANDY ANDY No. 2	
To multiply by	Simply
10, 100, 1000, etc.	move decimal point one, two, three, etc. places to the *right* in multiplicand.
0.1, 0.01, 0.001, etc.	move decimal point one, two, three, etc. places to the *left* in multiplicand.
5, 50, 500, etc.	multiply by 10, 100, 1000, etc. and divide by 2.
25, 250, etc.	multiply by 100, 1000, etc. and divide by 4.
125	multiply by 1000 and divide by 8.
$33\frac{1}{3}, 16\frac{2}{3}, 12\frac{1}{2}, 8\frac{1}{3}, 6\frac{1}{4}$	multiply by 100 and divide by 3, 6, 8, 12, 16.

Problems III.

1. Find the products.
    (a) 32×7.   (b) 326×8.   (c) 13.96×5.
    (d) 69.8×3.   (e) 6739×382.   (f) 46.37×146.
    (g) 75.93×8.32.   (h) 3.75×0.035.   (i) 0.132×0.047.
    (j) $\$38.75 \times 12$.   (k) $\$56.93 \times 0.045$.   (l) $\$1.32 \times 1238$.

2. If the railroad fare is computed at the rate of 3¢ a mile and the distance from New York to Chicago is 906 miles, how much is the fare?

3. A dealer orders 2 dozen sweaters valued at $5.76 per sweater and 3 dozen pairs of golf shoes valued at $11.98 per pair, and returns 24 dozen golf balls valued at 75¢ per ball. How much does he owe the distributor?

4. If the tax rate is $0.0625 per dollar of valuation and your property is valued at $8765.00, how much tax should you pay?

5. During World War II a certain type of fighter airplane cost the government about $93,000 each. If the government bought 269 per month for 5 months, how much did they cost?

6. "Buy this car for $200.00 down and $69.75 per month." If the payments run for 36 months, how much will you pay for the car?
7. "Price: 69¢ each, $7.78 per dozen." How much will you save if you buy them by the dozen instead of individually?
8. If your disability insurance pays $67.85 per week, how much would you collect for a disability lasting 19 weeks?
9. What is the cost of painting an area of 3270 square feet at 4.5¢ per square foot?
10. If there are 144 square inches in a square foot, how many square inches are there in 37.25 square feet?
11. If there are 23 accidents in the United States per minute, how many accidents will there be in 365 days? (There are 60 minutes in one hour and 24 hours in one day.)

Division

The process of division is the inverse of multiplication. We are now asked to find how many times one number "is contained" in another number. The number to be divided is called the *dividend;* the number by which we divide is called the *divisor;* the result is called the *quotient.* If a divisor is not contained an exact (integral) number of times in the dividend, the amount "left over" is called the *remainder.* Division is then defined as:

$$\frac{\text{Dividend}}{\text{Divisor}} = \text{Quotient} + \frac{\text{Remainder}}{\text{Divisor}}$$

or expressed in another way:

$$\text{Dividend} = (\text{Quotient} \times \text{Divisor}) + \text{Remainder}.$$

If the divisor contains only one digit, the division can be accomplished by the "short division" process.

Example

$$\text{Divisor} \quad 7)\overset{2}{794} \quad \text{Dividend}$$
$$113\tfrac{3}{7} \quad \text{Quotient}$$

The divisor, 7, is contained in the first digit of the dividend one time with no remainder. The number 1 is placed below the 7 of the dividend. The divisor, 7, is contained in the second digit, 9, one time with a remainder of 2. The 1 is placed below the 9, and the remainder, 2, is placed above and in front of the next digit 4 so that it reads 24. The divisor, 7, is contained in 24 three times with a remainder of 3. The answer is, therefore, $113\tfrac{3}{7}$.

If the divisor contains more than one digit, the division is accomplished by the "long division" process.

Example

```
                 7109    Quotient
Divisor   52)369675     Dividend
          364
          ---
           56
           52
           ---
          475
          468
          ---
            7    Remainder
```

The first step is to consider the number formed with the smallest number of digits in the dividend, beginning at the left, which contains the divisor and mentally calculate its divisibility by the divisor. Since 36 is less than 52, we consider first the number 369 which can be divided by 52 approximately 7 times. Therefore 7 is the first partial quotient and is placed above the 9 of the first trial dividend. The divisor, 52, is now multiplied by 7 and the product is subtracted from 369 to yield the remainder of 5. We now "carry down" the next digit of the dividend, 6, and place it to the right of the remainder 5 to form the number 56 which is the second trial dividend. The divisor, 52, is contained in the second trial dividend, 56, one time. The second partial quotient, 1, is placed to the right of the first partial quotient, 7, and directly above the carry down number, 6. The product of 52×1 is subtracted from 56 to yield a remainder of 4. The next digit of the dividend, 7, is carried down to form the third trial dividend, 47. Since 47 is less than the divisor, 52, the divisor is not contained in the third trial dividend and a zero is placed in the third trial quotient to the right of 71 and directly above the 7 of the dividend. The next digit of the dividend, 5, is now carried down to form the fourth trial dividend, 475. The divisor, 52, is contained in 475 nine times. The number 9 is placed in the quotient directly to the right of 710 and above the digit 5 in the dividend. The product, 52×9, is subtracted from the fourth trial dividend, 475, to yield a remainder of 7. Since we have used all the digits of the dividend, the answer is $7109\frac{7}{52}$.

Before we consider the division of a decimal number let us discuss the accuracy of a decimal number. For example, when we express the common fraction $\frac{1}{3}$ as a decimal (.3333, etc.), there is a question where the decimal fraction should stop. This is decided by a statement which specifies the desired number of significant digits.

The **significant digits** *of a decimal number are those beginning with the first one, reading from left to right, which is not zero and ending with the last one definitely specified.*

Examples

.3333 has four significant digits;
1.3962 has five significant digits;
0.00369 has three significant digits.

Decimal numbers can now be "rounded off" to a specified number of significant digits by the following rule:

To round a decimal number: (1) decide on the number of nonzero digits to be retained and discard all digits to the right of this; (2) if the digit to the immediate right of the last digit retained is greater than 5, increase by one the last digit retained; (3) if less than 5, leave the last digit unchanged.

Examples

36.247 to four significant digits is 36.25.
36.243 to four significant digits is 36.24.
0.00387625 to four significant digits is 0.003876.

If the number discarded is **exactly** 5, we have to adopt an arbitrary rule and most people use:

(a) if the number just preceding the 5 is *even*, leave it unaltered;
(b) if the number preceding the 5 is *odd*, increase it by one.*

Examples

32.245 to four significant digits is 32.24.
24.675 to four significant digits is 24.68.

Let us now consider division with decimal numbers.

Example

Obtain the quotient of 794 ÷ 7 correct to two decimal places.

7)794.000
113.428 or 113.43 *Ans.*

The division is the same as before except that we place zeros after the

---

* When dealing with money, financial institutions usually adopt an arbitrary rule which is to their advantage. If money is involved divisions are usually carried to 4 decimal places before rounding off to the nearest cent.

decimal point and continue the division far enough so that we can round off to two decimal places.

If the divisor is a decimal number, then the first step is to clear the decimal point from the divisor. This is accomplished by multiplying *both* the divisor and the dividend by 10 and repeating the operation as many times as required to make the divisor a whole number. The operation amounts to moving the decimal point to the right the *same* number of places in *both* the divisor and the dividend and annexing zeros to the dividend if necessary.

Example

Find 14.42 ÷ 0.013 to two decimals.

$$
\begin{array}{r}
1\ 109.23 \quad \text{Quotient} \\
\text{Divisor} \quad \ _\times 013.)\overline{14_\times 420.00} \quad \text{Dividend} \\
\underline{13} \\
1\ 4 \\
\underline{1\ 3} \\
120 \\
\underline{117} \\
3\ 0 \\
\underline{2\ 6} \\
40 \\
\underline{39} \\
1 \quad \text{Remainder}
\end{array}
$$

First the decimal point is moved 3 places to the right in the divisor *and* the dividend. Since $\frac{1}{13}$ is less than $\frac{1}{2}$, the next partial quotient is less than 5 and is discarded in an answer correct to two decimal places. To check a division, multiply the quotient by the divisor and add the remainder; the total should equal the dividend.

Example

$$
\begin{array}{r}
91.57 \\
36)\overline{3296.75} \\
\underline{324} \\
56 \\
\underline{36} \\
20\ 7 \\
\underline{18\ 0} \\
2\ 75 \\
\underline{2\ 52} \\
23
\end{array}
\qquad
\begin{array}{r}
\textit{Check:} \quad 91.57 \\
\times \quad\ 36 \\
\overline{549\ 42} \\
2747\ 1 \\
\overline{3296.52} \\
+\quad\ 23 \\
\overline{3296.75}
\end{array}
$$

HANDY ANDY No. 3

To divide by	Simply
10, 100, 1000, etc.	move decimal point one, two, three, etc. places to the *left* in dividend.
0.1, 0.01, 0.001, etc.	move decimal point one, two, three, etc. places to the *right* in dividend.
$3\frac{1}{3}$	multiply by 3 and divide by 10.
$33\frac{1}{3}$	multiply by 3 and divide by 100.
$333\frac{1}{3}$	multiply by 3 and divide by 1000.
$16\frac{2}{3}$	multiply by 6 and divide by 100.
$12\frac{1}{2}$	multiply by 8 and divide by 100.
$8\frac{1}{3}$	multiply by 12 and divide by 100.
25	multiply by 4 and divide by 100.
50	multiply by 2 and divide by 100.
125	multiply by 8 and divide by 1000.

A frequent application of division is that of finding an average. In mathematical language an average is the *arithmetic mean* and is found by adding the individual quantities and dividing by the number of quantities.

Example

If four automobile tires were bought at prices of $22.32, $26.84, $18.76, and $28.44, what was the average cost of a tire?

Solution. Add the costs and divide by 4:

```
        22.32
        26.84
        18.76
        28.44
  4 ) 96.36 | 24.09        $24.09  Ans.
```

Another type of average is one in which the addition has already been performed, such as average speeds.

Example

If it took $7\frac{1}{2}$ hours to travel 326 miles, what was the average speed per hour?

Solution.

$$43.46 \quad \text{or} \quad 43.5 \text{ mi./hr.} \quad Ans.$$

```
        43.46   or   43.5 mi./hr.  Ans.
  7.5)326.000
      300
      ───
      260
      225
      ───
      350
      300
      ───
      500
```

When performing a series of multiplications and divisions involving approximate numbers (numbers containing decimals and rounded off to a certain number of significant digits), we are not always sure of the last digit. Consider, for example, the problem

$$\frac{2(8)}{3.1416}.$$

This problem could be done in the following three ways:

(a) $\dfrac{(2)(8)}{3.1416} = \dfrac{16}{3.1416} = 5.0929.$

(b) $2\left(\dfrac{8}{3.1416}\right) = 2(2.5465) = 5.0930.$

(c) $(16)\left(\dfrac{1}{3.1416}\right) = 16(0.3183) = 5.0928.$

We note that the answers all differ in the last stated digit (they agree to 3 decimal places). These errors are known as rounding-off errors and are of great concern in the use of electronic calculators. The usual practice is to do the multiplying first and let the division be the last step, as was done in (a) above. It is also common practice to carry more digits than are required for all the intermediate steps and to round off in the last answer. Thus in (b) above we would write

$$2\left(\frac{8}{3.1416}\right) = 2(2.54647) = 5.09294 = 5.0929.$$

In (c) it would be necessary to write

$$16\left(\frac{1}{3.1416}\right) = 16(0.318309) = 5.092944 = 5.0929.$$

Problems IV.

1. Divide the following using short division:
   (a) $8632 \div 4.$
   (b) $9134 \div 7.$
   (c) $9264 \div 8.$
   (d) $13,426 \div 9.$
   (e) $64.93 \div 8.$
   (f) $3872 \div .6.$

2. Divide the following using long division and expressing the answer correct to 2 decimal places:
   (a) 168.31 ÷ 19.43. (b) .78 ÷ 3.56. (c) $1236.75 ÷ 12.
   (d) 96482 ÷ 1346. (e) 1,368,432 ÷ 468. (f) 3.692 ÷ 0.0036.

3. The balance due on a car after trade-in and down payment is $2675.00. If this is to be paid in 36 equal monthly payments, how much is each payment to the nearest cent?

4. A shipment of 14 pairs of the same shoes cost $122.50. How much did each pair cost?

5. A real estate subdivider wanted to obtain 15 lots in a block which was 975 feet long. How much frontage footing did each lot get?

6. The State of Texas has an area of 267,339 square miles. Rhode Island has an area of 1214 square miles. Into how many states of the size of Rhode Island could Texas be divided and how many square miles would be left over?

7. Each time the wheel of a car makes one complete turn (a revolution), the car travels a distance equal to the circumference of the wheel (assuming no slippage). If the circumference of a wheel is 7 feet and there are 5280 feet in a mile, how many revolutions does the wheel make when the car travels 248 miles?

8. If a Federal Agency bought 375 cars of a certain make and paid $853,443.75, what was the price of each car?

9. If there are 144 square inches in a square foot, how many square feet are there in 3888 square inches?

10. If a student had scores of 93, 87, 89, 98, and 79 on 5 examination papers in mathematics, what was his average score?

11. A car travels 298 miles in 6 hours 15 minutes. What is the average speed? (15 minutes = .25 hours.)

12. Find the average cost of 343 articles if the total bill for all of them was $87,368.00.

13. The average family car travels about 12,000 miles per year and gets about 16 miles to the gallon of gasoline. The oil changes average about 1 quart of oil for every 200 miles. If the cost of gasoline is 30¢ a gallon and oil is 55¢ a quart, how much is the annual cost for gas and oil?

14. At a special sale price, a man bought 3 shirts for $10.00. If he saved 89⅔¢ per shirt what was the original price per shirt?

15. A man made a trip at an average speed of 45 miles per hour (mph) and the return trip at 37 mph. What was the average for the round trip?

Operations with Common Fractions

The common fraction indicates a division. Thus $\frac{3}{4}$ is another way of saying $3 \div 4$. Consequently, any proper fraction can be expressed as a decimal fraction by performing the division.

Examples

$$\frac{3}{4} = .75, \quad \frac{1}{8} = .125, \quad \frac{2}{5} = .4.$$

Any improper fraction can be expressed as a mixed number by performing the division.

Examples

$$\frac{3}{2} = 1\frac{1}{2}, \quad \frac{15}{8} = 1\frac{7}{8}, \quad \frac{9}{4} = 2\frac{1}{4}.$$

Any mixed number can be expressed as a fraction by multiplying the whole number by the denominator and adding the numerator to the product to form the numerator of the resulting fraction. The denominator remains unchanged.

Examples

$$2\frac{1}{2} = \frac{(2 \times 2) + 1}{2} = \frac{5}{2}, \quad 3\frac{7}{8} = \frac{(3 \times 8) + 7}{8} = \frac{31}{8}.$$

The value of a fraction is unchanged if the numerator and denominator are multiplied by the same number or divided by the same number.

Examples

$$\frac{4}{5} = \frac{4 \times 3}{5 \times 3} = \frac{12}{15}; \quad \frac{15}{45} = \frac{15 \div 5}{45 \div 5} = \frac{3}{9} = \frac{3 \div 3}{9 \div 3} = \frac{1}{3}.$$

A fraction is in its **lowest terms** when the numerator and denominator are **prime** to each other. This is the same as saying they have no common factors; thus to reduce a fraction to its lowest terms we divide both numerator and denominator by their common factors.

Example

$$\frac{560}{630} = \frac{56}{63} = \frac{8}{9}.$$

First we divided both numerator and denominator by 10, then by 7.

Fractions which have the same denominator are called **similar fractions** *or fractions with a* **common denominator.**

If we are given a group of fractions we can always change them into similar fractions by finding a common denominator, but in so doing we desire to find the smallest such common denominator and this is called the **least common denominator** (l.c.d.). Usually the l.c.d. can be found by inspection, but for a large group having large denominators it is convenient to have a scheme for finding the l.c.d. The best scheme is to decompose each denominator into its factors and arrange them in columns. The l.c.d. is then found by taking one number from each column.

Example

Find the l.c.d. of $\frac{1}{30}$, $\frac{7}{25}$, $\frac{8}{45}$, and $\frac{12}{105}$.

$$
\begin{aligned}
30 &= 2 \times 3 \times 5 \\
25 &= \qquad\quad 5 \times 5 \\
45 &= \quad\; 3 \times 5 \qquad \times 3 \\
105 &= \quad\; 3 \times 5 \qquad\qquad \times 7 \\
\hline
\text{l.c.d.} &= 2 \times 3 \times 5 \times 5 \times 3 \times 7 = 3150.
\end{aligned}
$$

To change the given fractions to similar fractions we multiply each numerator by those factors in the l.c.d. which are not contained in the corresponding denominator. In the above example:

For $\frac{1}{30}$: $\quad 5 \times 3 \times 7 = 105$;
For $\frac{7}{25}$: $\quad 2 \times 3 \times 3 \times 7 = 126$;
For $\frac{8}{45}$: $\quad 2 \times 5 \times 7 = 70$;
For $\frac{12}{105}$: $2 \times 5 \times 3 = 30$.

Thus we have

$$\frac{1}{30} = \frac{1 \times 105}{30 \times 105} = \frac{105}{3150};$$

$$\frac{7}{25} = \frac{7 \times 126}{25 \times 126} = \frac{882}{3150};$$

$$\frac{8}{45} = \frac{8 \times 70}{45 \times 70} = \frac{560}{3150};$$

$$\frac{12}{105} = \frac{12 \times 30}{105 \times 30} = \frac{360}{3150}.$$

We shall now consider the four operations with common fractions.

I. Addition of Fractions.

To add proper fractions, first change them to fractions with an l.c.d., then add the resulting numerators and place the sum over the l.c.d.

24   *Arithmetic*

Example

Add $\frac{3}{8} + \frac{5}{6} + \frac{7}{12} + \frac{1}{15}$.

Solution.

$$8 = 2 \times 2 \times 2$$
$$6 = 2 \qquad\quad \times 3$$
$$12 = 2 \times 2 \qquad \times 3$$
$$15 = \qquad\qquad 3 \times 5$$
$$\text{l.c.d.} = 2 \times 2 \times 2 \times 3 \times 5 = 120.$$

Fraction	Numerator
$\frac{3}{8}$	$3 \times 15 = \ \ 45$
$\frac{5}{6}$	$5 \times 20 = 100$
$\frac{7}{12}$	$7 \times 10 = \ \ 70$
$\frac{1}{15}$	$1 \times \ \ 8 = \ \ \ \ 8$
	Sum $\ \ = 223$

Therefore the sum is $\frac{223}{120} = 1\frac{103}{120}$. *Ans.*

To add mixed numbers, change the fractional parts to fractions with an l.c.d., then add the whole numbers and fractions separately, and finally combine the two sums.

Example

Add $3\frac{5}{8} + 14\frac{3}{4} + 8\frac{1}{5} + 6\frac{7}{8}$.

Solution. The l.c.d. $= 40$.

$3\frac{5}{8} = 3$	$5 \times \ \ 5 = 25$
$14\frac{3}{4} = 14$	$3 \times 10 = 30$
$8\frac{1}{5} = 8$	$1 \times \ \ 8 = \ \ 8$
$6\frac{7}{8} = 6$	$7 \times \ \ 5 = 35$
31	98

$$31 + \frac{98}{40} = 31 + \frac{49}{20} = 31 + 2\frac{9}{20} = 33\frac{9}{20}. \ \ Ans.$$

II. Subtraction of Fractions.

To find the difference between two proper fractions, first change to fractions with an l.c.d., then subtract the resulting numerators and place the difference over the l.c.d.

Example

$$\frac{5}{6} - \frac{3}{4} = \frac{10 - 9}{12} = \frac{1}{12}.$$

To subtract a fraction from a whole number, subtract one unit from the whole number and convert it into a fraction with the same denominator as the subtrahend; then find the difference of the fractions.

Example

$$38 - \tfrac{17}{35} = 37\tfrac{35}{35} - \tfrac{17}{35} = 37 + \frac{35 - 17}{35} = 37\tfrac{18}{35}.$$

The difference of two mixed numbers can be found in three ways:

Example

Subtract $3\tfrac{2}{3}$ from $17\tfrac{1}{4}$.

Solution. The l.c.d. is 12.

(a) As mixed numbers:

$$17\tfrac{1}{4} = 17\tfrac{3}{12} = 16\tfrac{15}{12}$$
$$3\tfrac{2}{3} = 3\tfrac{8}{12} = 3\tfrac{8}{12}$$
$$\overline{\text{Difference} = 13\tfrac{7}{12}.}$$

(b) As improper fractions:

$$17\tfrac{1}{4} = \tfrac{69}{4} = \tfrac{207}{12}$$
$$3\tfrac{2}{3} = \tfrac{11}{3} = \tfrac{44}{12}$$
$$\overline{\text{Difference} = \tfrac{163}{12} = 13\tfrac{7}{12}.}$$

(c) As decimal fractions:

$$17\tfrac{1}{4} = 17.25$$
$$3\tfrac{2}{3} = 3.67$$
$$\overline{\text{Difference} = 13.58.}$$

III. Multiplication of Fractions.

To multiply a fraction by a fraction, multiply the numerators together to form the numerator of the product and multiply the denominators together to form the denominator of the product; reduce the product to its lowest terms.

Examples

(a) $\dfrac{3}{7} \times \dfrac{2}{5} = \dfrac{3 \times 2}{7 \times 5} = \dfrac{6}{35}.$

(b) $\dfrac{7}{9} \times \dfrac{3}{28} = \dfrac{7 \times 3}{9 \times 28} = \dfrac{21}{252} = \dfrac{1}{12}.$

In order to reduce the size of the numbers, divide the numerator and denominator by common factors wherever possible. This procedure is called *cancellation* of common factors.

Examples

(a) $\dfrac{\overset{1}{\cancel{7}}}{\underset{3}{\cancel{9}}} \times \dfrac{\overset{1}{\cancel{3}}}{\underset{4}{\cancel{28}}} = \dfrac{1 \times 1}{3 \times 4} = \dfrac{1}{12}.$

(b) $\dfrac{\overset{1}{\cancel{3}}}{\underset{2}{\cancel{4}}} \times \dfrac{\overset{1}{\cancel{2}}}{\underset{1}{\cancel{3}}} \times \dfrac{\overset{1}{\cancel{7}}}{9} \times \dfrac{5}{\underset{2}{\cancel{14}}} = \dfrac{1 \times 1 \times 1 \times 5}{2 \times 1 \times 9 \times 2} = \dfrac{5}{36}.$

To multiply mixed numbers, change them to improper fractions and multiply as with fractions.

Example

$$2\tfrac{3}{4} \times 6\tfrac{4}{11} = \tfrac{11}{4} \times \tfrac{70}{11} = \tfrac{35}{2} = 17\tfrac{1}{2}.$$

IV. Division of Fractions.

Any number can be written as that number over one; i.e., 3 can be written as $\tfrac{3}{1}$. The number $\tfrac{1}{3}$ can be obtained from $\tfrac{3}{1}$ by changing the position of the two integers 1 and 3. This process is called *inverting*. Thus if we invert 5 we obtain $\tfrac{1}{5}$; inverting $\tfrac{3}{4}$ yields $\tfrac{4}{3}$, etc. Since division is the inverse of multiplication, we see that dividing by 3 and multiplying by $\tfrac{1}{3}$ yields the same answer. Thus to divide a number by another, we can obtain the answer by inverting the divisor and multiplying the dividend by the resulting number. This fact is used when dividing a fraction by another fraction and we state the following rule:

To divide a fraction by a fraction, invert the fraction of the divisor and multiply the result by the dividend.

Examples

(a) $\tfrac{7}{12} \div \tfrac{7}{4} = \tfrac{7}{12} \times \tfrac{4}{7} = \tfrac{1}{3}.$
(b) $\tfrac{3}{4} \div 6 = \tfrac{3}{4} \times \tfrac{1}{6} = \tfrac{1}{8}.$
(c) $12 \div \tfrac{2}{5} = \tfrac{12}{1} \times \tfrac{5}{2} = 30.$

Division of mixed numbers is accomplished by changing the mixed numbers to improper fractions and then applying the rule for division of fractions.

Examples

(a) $2\tfrac{5}{8} \div 4\tfrac{2}{3} = \tfrac{21}{8} \div \tfrac{14}{3} = \tfrac{21}{8} \times \tfrac{3}{14} = \tfrac{9}{16}.$
(b) $7\tfrac{2}{9} \div 3\tfrac{3}{4} = \tfrac{65}{9} \div \tfrac{15}{4} = \tfrac{65}{9} \times \tfrac{4}{15} = \tfrac{52}{27}.$
(c) $2\tfrac{1}{4} \div 3 = \tfrac{9}{4} \div 3 = \tfrac{9}{4} \times \tfrac{1}{3} = \tfrac{3}{4}.$

HANDY ANDY No. 4

Decimal Equivalents of Common Fractions

$\frac{1}{64} = .015625$	$\frac{11}{32} = .34375$	$\frac{11}{16} = .6875$
$\frac{1}{32} = .03125$	$\frac{3}{8} = .375$	$\frac{23}{32} = .71875$
$\frac{1}{16} = .0625$	$\frac{13}{32} = .40625$	$\frac{3}{4} = .75$
$\frac{3}{32} = .09375$	$\frac{7}{16} = .4375$	$\frac{25}{32} = .78125$
$\frac{1}{8} = .125$	$\frac{15}{32} = .46875$	$\frac{13}{16} = .8125$
$\frac{5}{32} = .15625$	$\frac{1}{2} = .5$	$\frac{27}{32} = .84375$
$\frac{3}{16} = .1875$	$\frac{17}{32} = .53125$	$\frac{7}{8} = .875$
$\frac{7}{32} = .21875$	$\frac{9}{16} = .5625$	$\frac{29}{32} = .90625$
$\frac{1}{4} = .25$	$\frac{19}{32} = .59375$	$\frac{15}{16} = .9375$
$\frac{9}{32} = .28125$	$\frac{5}{8} = .625$	$\frac{31}{32} = .96875$
$\frac{5}{16} = .3125$	$\frac{21}{32} = .65625$	$\frac{63}{64} = .984375$

Problems V.

1. Reduce the following fractions to their lowest terms:
   (a) $\frac{24}{30}$.   (b) $\frac{315}{378}$.   (c) $\frac{1890}{7560}$.   (d) $\frac{4.62}{1.98}$.   (e) $\frac{55440}{147840}$.

2. Add the following fractions:
   (a) $\frac{1}{2} + \frac{1}{3} + \frac{1}{4}$.   (b) $\frac{1}{2} + \frac{2}{3} + \frac{3}{4} + \frac{4}{5} + \frac{5}{6}$.
   (c) $\frac{7}{12} + \frac{5}{36} + \frac{11}{24}$.   (d) $\frac{8}{12} + \frac{21}{28} + \frac{1}{6}$.
   (e) $2\frac{1}{2} + 3\frac{1}{3}$.   (f) $12\frac{3}{4} + 2\frac{7}{16} + 3\frac{3}{8}$.

3. Subtract the following fractions:
   (a) $\frac{1}{2} - \frac{1}{3}$.   (b) $\frac{17}{12} - \frac{5}{4}$.   (c) $\frac{52}{121} - \frac{7}{33}$.
   (d) $13 - \frac{7}{9}$.   (e) $7\frac{2}{5} - 3\frac{1}{3}$.   (f) $9\frac{1}{16} - 3$.

4. Multiply the following fractions:
   (a) $\frac{2}{3} \times \frac{3}{5}$.   (b) $\frac{7}{9} \times \frac{81}{56}$.   (c) $\frac{32}{45} \times \frac{25}{96}$.
   (d) $\frac{12}{35} \times \frac{25}{96} \times \frac{49}{100}$.   (e) $\frac{21}{25} \times \frac{5}{28} \times \frac{8}{3}$.
   (f) $2\frac{1}{3} \times 3\frac{1}{2}$.   (g) $7\frac{3}{5} \times 9\frac{1}{3}$.   (h) $4\frac{1}{5} \times 3\frac{4}{7} \times \frac{2}{15}$.

5. Divide the following fractions:
   (a) $\frac{1}{2} \div \frac{1}{3}$.   (b) $\frac{7}{11} \div \frac{3}{11}$.   (c) $\frac{21}{75} \div \frac{7}{25}$.
   (d) $2\frac{11}{27} \div \frac{5}{9}$.   (e) $6\frac{3}{4} \div 3$.   (f) $13\frac{59}{75} \div 3\frac{1}{5}$.

6. Perform the indicated operations:
   (a) $\frac{7}{12} + \frac{15}{4} - \frac{5}{6}$.   (b) $(\frac{2}{3} + \frac{1}{2}) \times \frac{3}{14}$.
   (c) $(\frac{3}{2} \times \frac{1}{5}) \div \frac{1}{10}$.   (d) $(\frac{3}{7} - \frac{1}{21}) \times \frac{7}{16}$.

7. A steel rod was cut into five pieces of lengths $6\frac{1}{4}$, $3\frac{1}{2}$, $2\frac{11}{16}$, $3\frac{9}{16}$, and $7\frac{3}{4}$ inches, respectively. How long was the rod if $\frac{1}{16}$ inch was wasted in each cut?

8. If 7 is added to the numerator and also to the denominator of $\frac{1}{3}$, is the value of the fraction increased or diminished, and by what amount?

9. If the gas tank of an automobile holds 21 gallons, how many gallons are in the tank if the indicator shows $\frac{3}{4}$ full? If the indicator shows $\frac{1}{4}$ full?

10. You are traveling by car on a vacation trip and average $37\frac{8}{10}$ miles per hour and $8\frac{2}{3}$ hours per day. How far have you traveled at the end of $3\frac{1}{2}$ days?

11. Let us assume that a city block averages $\frac{1}{8}$ mile. How far have you walked in going 7 blocks west and $5\frac{1}{2}$ blocks south?

12. A recipe calls for ingredients in portions of 4 cups, $1\frac{1}{2}$ cups, $\frac{2}{3}$ cup, 2 cups, and $\frac{1}{8}$ cup (actually 2 tablespoons). How big a bowl should you have? (How many cups must the bowl hold?)

13. A man whose wage rate is $2.18 per hour is paid $1\frac{1}{2}$ times that rate for working overtime. If he works $5\frac{1}{4}$ hours overtime, what are his overtime earnings? What would be his pay if he worked only $38\frac{1}{4}$ hours in one week and his regular working week is 40 hours?

Operations with Negative Numbers

We defined a negative number on p. 5 and used a minus sign, "$-$," to indicate that a number was negative. It is now important to realize that the plus sign, $+$, and the minus sign, $-$, each means one of two things; i.e., a plus sign may

        (a) indicate an operation (to add),
        (b) indicate a positive number;

a minus sign may

        (a) indicate an operation (to subtract),
        (b) indicate a negative number.

A little practice with these signs will soon indicate that the meanings are usually very clear and no confusion need arise.

Examples

(a) $+7$ or 7 denotes a positive number 7.
(b) -7 denotes a negative number 7.
(c) $3 + 5$ denotes the addition of a positive 3 and a positive 5.
(d) $3 + (-5)$ denotes the addition of a positive 3 and a negative 5.
(e) $7 - (-3)$ denotes the subtraction of a negative 3 from a positive 7.

We shall now turn to the four operations applied to both positive and negative numbers.

I. Addition with Negative Numbers.

Rule 1. *To* add *two or more numbers having* like signs, *find the sum of their numerical values and prefix the common sign before the result.*

Examples

(a) The sum of +5 and +2 is +7.
(b) The sum of −3 and −12 is −15.
(c) $(-2) + (-5) + (-7) = -14$.

Rule 2. *To* add *two numbers having* unlike signs, *find the difference of their numerical values and prefix to this result the sign of the number which has the greater numerical value.*

Examples

(a) The sum of +7 and −3 is +4.
(b) The sum of +7 and −15 is −8.
(c) $14 + (-10) = 4$.

II. Subtraction with Negative Numbers.

With the use of negative numbers, subtraction of a positive number now becomes the addition of a negative number. Thus, to subtract 5 from 7 and to add −5 and +7 is the same operation; the result is +2 in both cases. The use of negative numbers makes it possible to subtract a larger number from a smaller number. For example, to subtract 7 from 5 simply add −7 and +5 and by Rule 2 the answer is −2.

Rule 3. *To* subtract a negative number, *simply change its sign and add.*

When adding remember the two rules for addition.

Examples

(a) Subtract −3 from +7:
$$7 - (-3) = 7 + 3 = 10. \quad Ans.$$

(b) Subtract −3 from −7:
$$-7 - (-3) = -7 + 3 = -4. \quad Ans.$$

(c) $-\dfrac{3}{4} + \dfrac{4}{9} + \dfrac{3}{6} + \left(-\dfrac{1}{2}\right) = \dfrac{-27 + 16 + 18 - 18}{36} = -\dfrac{11}{36}. \quad Ans.$

(d) $\dfrac{1}{2} - \left(-\dfrac{2}{3}\right) + \dfrac{1}{6} - \dfrac{7}{12} = \dfrac{6 + 8 + 2 - 7}{12} = \dfrac{9}{12} = \dfrac{3}{4}. \quad Ans.$

III. Multiplication and Division with Negative Numbers.

Since multiplication and division are inverse operations we shall consider both operations at the same time. To multiply or divide signed numbers, first determine the sign of the result, then perform the indicated operation on the numerical values of the numbers.

Rule 4.

(a) *If the multiplier and the multiplicand have the* **same sign,** *the product is* **positive.**

(b) *If the multiplier and the multiplicand have* **opposite signs,** *the product is* **negative.**

(c) *If the dividend and the divisor have the* **same sign,** *the quotient is* **positive.**

(d) *If the dividend and the divisor have* **opposite signs,** *the quotient is* **negative.**

These rules can be put into a table for easy memorization.

Multiplication	Division
$+ \times + = +$	$+ \div + = +$
$+ \times - = -$	$+ \div - = -$
$- \times + = -$	$- \div + = -$
$- \times - = +$	$- \div - = +$

Examples

(a) $7 \times 5 = 35.$

(b) $32 \times (-5) = -160.$

(c) $(-12) \times (-3) = 36.$

(d) $8 \div 4 = 2.$

(e) $(-12) \div 3 = -4.$

(f) $(-36) \div (-6) = 6.$

If a multiplication problem contains more than two factors, the sign of the answer can be determined by counting the number of *negative signs* appearing in the problem; if the number is **odd,** the answer is **negative;** if the number is **even,** the answer is **positive.**

Examples

(a) $\left(-\frac{3}{2}\right) \times \left(\frac{4}{9}\right) \times \left(-\frac{1}{2}\right) = \frac{3}{2} \times \frac{4}{9} \times \frac{1}{2} = \frac{1}{3}.$ *Ans.*

(b) $\dfrac{(-6)(8)}{(2)(-5)} = \dfrac{-48}{-10} = \dfrac{24}{5} = 4\frac{4}{5}.$ *Ans.*

(c) $\dfrac{1}{5} \times \left(-\dfrac{6}{7}\right) \div \dfrac{-3}{14} = \left(\dfrac{1}{5}\right)\left(\dfrac{6}{7}\right)\left(\dfrac{14}{3}\right) = \dfrac{4}{5}.$ *Ans.*

(d) $\dfrac{3(-5)\left(-\frac{2}{3}\right)}{-\frac{5}{3}} \div \dfrac{3}{4} = -10 \times \dfrac{3}{5} \times \dfrac{4}{3} = -8.$ *Ans.*

(e) The thermometer drops from 26 degrees to −5 degrees. How many degrees did it drop?

Solution. $26 - (-5) = 31$ degrees. *Ans.*

Problems VI.

1. Perform the indicated operation.
   (a) $-.3 + 1.2.$
   (b) $-3 + 2 - (-1.5) + (-.75).$
   (c) $-1.69 - (-.97).$
   (d) $-\frac{1}{2} + (-\frac{2}{7}) - (-\frac{3}{8}) + \frac{5}{14}.$
2. Perform the indicated multiplication.
   (a) $(-5.2) \times (3.9).$
   (b) $(-\frac{1}{5})(\frac{5}{6})(-\frac{2}{3}).$
   (c) $(-.0123)(-.456)$ to four significant digits.
3. Perform the indicated division.
   (a) $-64.736 \div 35.92$ to four significant digits.
   (b) $(-\frac{3}{5}) \div (-\frac{6}{7}).$
   (c) $32 \div (-\frac{1}{2}).$
4. Perform the indicated operations.

$$(2\frac{3}{4}) \times (-1\frac{2}{3}) \div (-2\frac{1}{2})$$

The Number Zero

The number zero is a special number in our number system. In the first place it is both negative and positive, or we might say it is neither negative nor positive; in other words, it has no sign connected with it but is the dividing point between the negative and positive numbers. The operations with zero can be summarized as follows:

(a) Any number *plus* zero equals the number.

(b) Any number *minus* zero equals the number.

(c) Zero *minus* any number equals the negative of the number.

(d) Any number *times* zero equals zero.

(e) Zero *divided* by any number except zero equals zero.

(f) The operation of *dividing by* zero is **not defined** and is not permitted.

The last statement is the result of the fact that we cannot arrive at a definition for dividing by zero which is consistent with the other definitions that we have adopted for our number system.*

---

* In advanced mathematics we consider the problem of dividing a number (not zero) by a number which gets smaller and smaller; e.g., $3 \div \frac{1}{2} = 6$, then $3 \div \frac{1}{3} = 9$, then $3 \div \frac{1}{4} = 12$, $3 \div \frac{1}{10} = 30$, etc. We see that as the denominator gets smaller and smaller ($\frac{1}{2}, \frac{1}{3}, \frac{1}{4}, \frac{1}{10}$, etc.) the quotient gets larger and larger (6, 9, 12, 30, etc.). Now as the denominator approaches 0, the quotient approaches a number which is larger than any previously given number no matter how large. In the limit we say that the quotient approaches *infinity*.

"Casting Out Nines"

In Handy Andy No. 5 we give a brief outline of a method for checking long arithmetic problems. This method has long been used by
accountants. It is based on the idea of finding the sum of the digits
in a number and then adding the digits in the resulting sum, etc. until

---

HANDY ANDY No. 5

"Casting Out Nines"

Add the digits of a number discarding any whose sum is 9. If the
sum has more than one digit, add these and continue until a number of one digit is found. Call this addition "finding the K."

Example: $7, 346, 283 \rightarrow 15 \rightarrow 6$.

Check for addition:

Find the K of each number.
Add these K's.
Find the K of this total, K_T.
Find the K of the sum, K_S.
If $K_T = K_S$, addition is correct.

Example

Add	K
346925	2
673241	5
397163	2
367142	5
793241	8
2577712	22

check

$K_S = 4 \longleftrightarrow K_T = 4$

Check for multiplication:

Find the K for multiplicand, K_M.
Find the K for multiplier, K_m.
Find the K for the product, K_p.
If $K_M \times K_m = K_P$, answer is
correct.

Example

$36725 \qquad K_M = 5$
$936 \qquad K_m = 0$
$\overline{220350} \quad K_M \times K_m = 0$
110175
330525
$\overline{34374600}$

$K_P = 18 = 9 = 0$

Check for division:

Find the K for the dividend, K_D; the divisor, K_d; the quotient,
K_Q; and the remainder K_r. Then

$$(K_d \times K_Q) + K_r = K_D$$

a one digit number results. This addition of the digits in a number is further simplified by first discarding (throwing away) any digits whose sum is 9; for example, in the number 7,346,258,355, we discard the 7 and 2, 3 and 6, 4 and 5 and consider only 8355. Adding these digits gives a first sum of 21 and adding these digits gives 3, which we shall call K for convenience. The procedure for checking is shown in Handy Andy No. 5.

Review Exercises I.

1. Add:

(a)	(b)	(c)
42.396	$ 3.75	12.39
39.867	12.32	6.41
42.160	6.87	7.68
0.936	1.93	3.72
2.732	4.69	4.10
17.639	2.73	5.95
18.416	14.38	6.05

   (d) $\frac{1}{2} + \frac{1}{3} + \frac{3}{8} + \frac{11}{6}$.
   (e) $\frac{3}{8} + \frac{5}{7} + (-\frac{7}{12})$.

2. Perform the indicated operation:
   (a) $(46.12) \times (3.1416)$.
   (b) $(38.67) \times (-2.39)$.
   (c) $\frac{1}{2} \times \frac{4}{25} \div (-\frac{8}{5})$.
   (d) $96.312 \div (-31.62)$ to 3 decimal places.

3. Starting on a second floor landing with steps going both up and down, John went up 3 steps, down 5 steps, up 7 steps, down 11 steps, and up 6 steps. Where did he finish?

4. Mr. Jones had 5 cars which he had bought for $1235.00, $960.00, $2250.00, $400.00, $1875.00. He sold them all for $7500.00. How much was his profit?

5. If it costs 7¢ for the first pound and 1¢ for each additional pound to send a package by parcel post, how much would it cost to send a 12 pound package?

6. If it costs 50¢ for the first 10 words and 2¢ for each additional word to send a telegram, how much would it cost to send a 32 word telegram?

7. A receipt from a department store reads:

Qt.	Item		Cost
3	prs. hose at	0.89	$ 2.67
2	shirts at	3.78	7.56
1	pr. slacks		8.98
4	prs. socks at	0.36	1.54
			$22.85

Is the bill correct? If not, find the amount of the correct bill.

8. In 1960 John bought a car listed for $2876.50. He was allowed $1135.00 for his old car. John then had the dealer put in 10 gallons of gas at 29.9¢ a gallon and $3.75 worth of anti-freeze; the license cost $11.25 and the title $2.50. How much cash did John pay?

9. If oranges cost 72¢ a dozen, what is the cost of 9 oranges?

10. Tom wants to buy a bicycle which cost $39.75. He has saved $33.00 and gets an allowance of $1.25 a week of which he can save $0.75. How long will it take him to save enough money to buy the bicycle?

11. The temperatures at 3 A.M. during 7 days of January were 18, 9, 7, −8, −5, −3, and −11. What was the average temperature?

12. Find the following values:

(a)	(b)
$1 \times 8 + 1 =$	$0 \times 9 + 8 =$
$12 \times 8 + 2 =$	$9 \times 9 + 7 =$
$123 \times 8 + 3 =$	$98 \times 9 + 6 =$
$1234 \times 8 + 4 =$	$987 \times 9 + 5 =$
$12345 \times 8 + 5 =$	$9876 \times 9 + 4 =$
$123456 \times 8 + 6 =$	$98765 \times 9 + 3 =$
$1234567 \times 8 + 7 =$	$987654 \times 9 + 2 =$
$12345678 \times 8 + 8 =$	$9876543 \times 9 + 1 =$
$123456789 \times 8 + 9 =$	$98765432 \times 9 + 0 =$
	$987654321 \times 9 - 1 =$
	$9876543210 \times 9 - 2 =$

13. For the same volume ice is about $\frac{9}{10}$ as heavy as water. If water weighs $62\frac{1}{2}$ lbs. per cubic foot, how much do 5 cubic feet of ice weigh?

14. A streamlined train makes a run of 546 miles in 5 hours 36 minutes (= 5.6 hours). What is its average speed per hour?

15. If the price of gasoline is 31.9¢ per gallon and 10¢ of this is for taxes, how much does the customer pay on a purchase of 14.7 gallons? How much is the tax and how much does the dealer get?

Chapter II

ALGEBRA

Introduction

The subject of algebra is an extension of arithmetic. We shall again consider numbers and the operations with numbers. The basic concept which we shall add to our previous knowledge is the idea of a *general number;* that is, the representation of numbers by *letters.* For example, we may say a room is x feet long and x may stand for *any* number. If we are speaking of a particular room and measure it to be 10 feet long, then *in speaking of that particular room x* would equal 10; however, for another room x may have a different value. This is the idea of a general number and we use the letters of the alphabet in our mathematical representation. We may then think of algebra as the arithmetic of general numbers plus some new concepts such as formulas and equations. All of the knowledge we have learned in arithmetic will again be used so that it is important to remember what we have already learned.

Algebraic Symbols

Let us start by learning the meaning of some of the words and symbols of algebra. The signs of operation $(+, -, \times, \div)$ will continue to have the same meaning as they did in arithmetic. In algebra we shall use additional conventions, especially for multiplication. Thus by simply writing ab we shall mean a times b; by writing $3axy$ we shall mean 3 times a times x times y. We shall also use parentheses to indicate multiplication; thus $(a + b)(x - y)$ means $(a + b)$ times $(x - y)$. And, of course, we shall use the equal sign, $=$, which says "is equal to" or "is the same as."

In general, a combination of symbols and signs of algebra represents a number and is called an **algebraic expression**.

Example

$$3axy + 7bxz - 5c.$$

The part of the algebraic expression which is not separated by a plus or minus sign is called a **term**. In the above example, there are three terms: $3axy$, $+7bxz$, $-5c$. Note that we include the sign in the term although frequently we consider the plus sign to be understood. If we write a term without a number, axy, it is understood that we have one of these; for this reason the number 1 is omitted.

In an algebraic term, $3axy$, the parts 3, a, x, and y are called **factors** of the term. The product of all of them except one is called the **coefficient** of that one.

Examples

In $3axy$:

$$3 \text{ is the coefficient of } axy,$$
$$3a \text{ is the coefficient of } xy,$$
$$3ax \text{ is the coefficient of } y,$$
$$3y \text{ is the coefficient of } ax, \text{ etc.}$$

The numerical part is called the **numerical coefficient.** In the above example, 3 is the numerical coefficient.

In the representation of numbers by letters we may use a **subscript,** which is a small number slightly below and to the right of the letter (t_1). This number is *not* to be regarded numerically; it is simply a method to distinguish between various quantities which we are denoting by the same letter. For example, T_1 and T_2 may represent two *different* temperatures.

Algebraic expressions are given convenient names according to the number of terms. We summarize the definitions in the following table:

Number of Terms	Name	Example
1	monomial	$3ax$
2	binomial	$x + 2y$
3	trinomial	$3a + b - c$
2 or more	polynomial (or multinomial)	$x + y - c + 3$

Terms which are exactly the same except for their numerical coefficient are called **like terms;** otherwise they are called **unlike terms.** Thus $3ax$ and $4ax$ are like terms, and $3ax$ and $4ay$ are unlike terms.

In adding and subtracting algebraic terms only like terms may be combined. The sum and difference of unlike terms can only be expressed by the use of plus and minus signs.

Examples

$$3ay + 4ay - 2ay = 5ay.$$
$$3x + 2y - x = 2x + 2y.$$

In order to have a better understanding of the language of algebra we should be able to restate verbal expressions in terms of algebraic symbols.

Examples

Verbal Expression	Algebraic Expression
Three times a number	$3N$
A number increased by 7	$n + 7$
A number decreased by 12	$N - 12$
Twice a number increased by 5	$2x + 5$
The sum of three numbers	$n_1 + n_2 + n_3$
The sum of two consecutive integers	$x + (x + 1)$
The product of any three numbers	xyz
An even number	$2n$
An odd number	$2n + 1$
Twice the sum of two numbers	$2(a + b)$
The area equals base times width	$A = bw$
One fourth the width	$\frac{1}{4} w$ or $\frac{w}{4}$

Note that we may choose any letter to represent a quantity; however, some letters have become associated with certain expressions by convention. For example, A is used for area, h for height, etc.

In dealing with algebraic expressions it is often necessary to group together several parts. This is accomplished by using *parentheses*, ( ); *brackets*, [ ], and *braces*, { }. These symbols of grouping may be removed or inserted according to the following rules:

Rule 1. *Symbols of grouping preceded by a* **plus** *sign may be removed by rewriting each of the enclosed terms with its original sign.*

Example

$$3a + (2b - 5c) = 3a + 2b - 5c.$$

Rule 2. *Symbols of grouping preceded by a* **minus** *sign may be removed by rewriting and changing the sign of each of the enclosed terms.*

Example

$$3a - (2b - 5c) = 3a - 2b + 5c.$$

Rule 3. *If a* **coefficient** *precedes a symbol of grouping, it is to be* **multiplied** *into each included term when the symbol is removed.*

Example

$$x - 2(3y - 2a) = x - 6y + 4a.$$

To simplify expressions involving several symbols of grouping we work from the inside out by *first removing the innermost pair* of symbols, next the innermost pair of remaining ones, and so on. If like terms appear they are combined at each step.

Example

$$
\begin{aligned}
4a - b - \{3a - [2a(4 - b) - (a - b)]\} \\
= 4a - b - \{3a - [8a - 2ab - a + b]\} \\
= 4a - b - \{3a - 7a + 2ab - b\} \\
= 4a - b + 4a - 2ab + b \\
= 8a - 2ab.
\end{aligned}
$$

Powers and Roots

The product of like factors is conveniently expressed by symbols to form a shorthand method of writing. Thus, for example, $x \cdot x \cdot x$ is written as x^3. The product is called the **power** of the factors; i.e., x^3 is the *third power* of x. The number x is called the **base** and the small number 3 (written to the right and partially above x) is called the **exponent** of the power. The exponent is the number of times the factor appears in multiplication.

Examples

$$\underbrace{x \cdot x \cdot x \cdot x \cdot x}_{5 \text{ factors}} = x^5.$$

$$2 \cdot 2 \cdot 2 = 2^3 = 8.$$

We speak of x^2 as x square; x^3 as x cube, x^n as the nth power of x. It is convenient to know the small powers of the digits, and some of them should be memorized. Table I in the back of this book lists the squares and cubes of the numbers from 1 to 100.

HANDY ANDY No. 6

Powers of Small Integers

Power\Number	2	3	4	5	6	7	8	9
2	4	9	16	25	36	49	64	81
3	8	27	64	125	216	343	512	729
4	16	81	256	625	1296	2401	4096	6561
5	32	243	1024	3125	7776			
6	64	729	4096					
7	128							

To find 10 to a power simply annex to 1 the same number of zeros as the exponent: $10^2 = 100$; $10^6 = 1,000,000$; etc.

To square a number ending in 5, multiply the part preceding 5 by one plus that number and annex 25; e.g., $25^2 = [2 \times 3]25 = 625$; $65^2 = [6 \times 7]25 = 4225$.

We have already seen that in adding and subtracting algebraic terms we can combine only like terms. It is therefore important to note that x^2 and x^3 are *not* like terms and thus we can only express an addition or subtraction. Before we can multiply or divide terms which involve exponents we need some rules. These are known as the **basic laws of exponents.**

I. $a^m a^n = a^{m+n}$.

II. $(a^m)^n = a^{mn}$.

III. $(ab)^n = a^n b^n$.

IV. $\left(\dfrac{a}{b}\right)^n = \dfrac{a^n}{b^n}$.

V. $\dfrac{a^m}{a^n} = a^{m-n}$.

Examples

$2^3 \cdot 2^5 = 2^8 = 256$.

$(3^2)^3 = 3^6 = 729$.

$(2 \cdot 3)^2 = 2^2 \cdot 3^2 = 4 \cdot 9 = 36$.

$\left(\dfrac{2}{3}\right)^2 = \dfrac{2^2}{3^2} = \dfrac{4}{9}$.

$\dfrac{2^5}{2^2} = 2^3 = 8$.

We shall define $a^0 = 1$ providing a is not zero; that is, any quantity raised to the zero power is 1.

Examples

$$\frac{2^2}{2^2} = 2^0 = 1; \quad \frac{x^4}{x^4} = x^0 = 1.$$

If $a^m = N$, then a is the mth root of N. That is, we are looking for a number a such that when multiplied by itself to m factors the result is N. We write this statement with the use of a **radical** sign, $\sqrt{}$ thus:

$$\sqrt[m]{N} = a.$$

The number m is called the **index of the root** and N, the **radicand**.

If $m = 2$, we call it a *square root;* if $m = 3$, a *cube root.* The process of taking a root does not necessarily lead to a single answer. For example, the square root of 4 may be either $+2$ or -2 since both $(+2)(+2) = 4$ and $(-2)(-2) = 4$. To express this condition we introduce the symbol $\pm$, which reads *plus or minus*. The double root is necessary only when the index is an even number.

Examples

$$\sqrt{16} = \pm 4; \quad \sqrt[4]{81} = \pm 3;$$
$$\sqrt[3]{-8} = -2; \quad \sqrt[5]{32} = 2.$$

The square root of a number is usually not an integer. Since the square root of 9 is 3 and the square root of 16 is 4, then the square root of any number between 9 and 16, say 12, is between 3 and 4 and must be expressed by a decimal number, $3.+$. The easiest way to find the square root of a number is to use a table of roots (see Table I in the back of this book). Such tables are usually read by locating the number in the left column and its square root in this row and the column headed by "square root."

Example

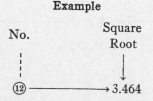

The roots of numbers can also be found by logarithms, which will be explained in Chapter VII.

The "long-hand" method of extracting the square root is not too difficult if a few simple rules are followed. The first rule is to determine the number of digits in the root; this is done by separating the number in pairs of two digits each, beginning at the right. The roots will contain one digit for each pair in the number. If an extra, single digit remains in the number, the roots will have one additional digit. For example: Take the number 2468. Separating into pairs of two digits each, we have 24 68 or two pairs. Hence, the roots will consist of two digits. If the number is 3 24 68, we have two pairs and one extra digit; hence, the roots will consist of three digits.

Suppose the number contains a decimal, say, 32468.296. We follow the same procedure, except that we separate the number into pairs, working to the left and to the right of the decimal point. Complete pairs must be formed to the right of the decimal point. If an extra digit remains, a zero must be added to form an even pair. Thus, separating the above number into pairs, we have 3 24 68.29 60. We can add as many pairs of zeros as we deem necessary in order to obtain the roots to the required degree of accuracy. The decimal point in the number is carried in column to the root obtained. An example is worked out below, and the operations involved are explained in detail. Extract the square root of 32468.296. Place the number in the radical sign $\sqrt{}$ and separate the number into pairs.

$$
\begin{array}{r}
1\ \ 8\ \ 0.\ 1\ \ 8\ \ 9 \\
\sqrt{3\ 24\ 68.29\ 60\ 00}
\end{array}
$$

8	1
2Ø	2 24
1	2 24
360Ø	0 00 68 29
8	36 01
3602Ø	32 28 60
9	28 82 24
36036Ø	3 46 36 00
	3 24 33 21
	22 02 79 Remainder

Procedure. The largest square contained in the first (extra) digit, 3, is 1. The 1 is the first partial root, and is placed above the 3. The first partial root is squared and subtracted from 3. Bring down the next pair to the right of the difference, forming a new number, 224. The first partial root, 1, is doubled and brought down to the left of the new number, 224. Temporarily, annex a zero to the 2, making the first trial divisor 20, and then mentally calculate the number which can be substituted for the 0 and will give a product not greater than 224. This number, which works out to 8, is the second partial root, and is placed above the 24. The product of 28 by 8 is 224, which is subtracted from the first difference to obtain the second difference which in this case is 0. The two partial roots, so far obtained, form a number 18, which is doubled and carried to the left of the second difference. A zero is again temporarily annexed to the 36, making the second trial divisor 360. As 360 is greater than 68, the third partial root is 0. The next pair is carried down and the above operations are repeated.

The roots of numbers can also be expressed by the use of fractional exponents. That is, the square root of x is the same as $x^{\frac{1}{2}}$; cube root of x, $x^{\frac{1}{3}}$, etc. In general we have

$$\sqrt[n]{x} = x^{\frac{1}{n}}.$$

To complete our definitions of exponents, we shall define the negative exponents as follows:

$$a^{-n} = \frac{1}{a^n}.$$

Thus the negative exponents indicate the reciprocal of the quantity raised to the same positive exponent.

Examples

$$3^{-2} = \frac{1}{3^2} = \frac{1}{9}; \quad \frac{1}{2^3} = 2^{-3};$$
$$(2^{\frac{1}{2}})(2^{\frac{1}{2}}) = 2^{\frac{1}{2}+\frac{1}{2}} = 2^1 = 2.$$

HANDY ANDY No. 7	
Powers of 10	
$10^1 = 10$	$10^{-1} = 0.1$
$10^2 = 100$	$10^{-2} = 0.01$
$10^3 = 1000$	$10^{-3} = 0.001$
$10^4 = 10,000$	$10^{-4} = 0.0001$
$10^5 = 100,000$	$10^{-5} = 0.00001$
$10^6 = 1,000,000$	$10^{-6} = 0.000001$
etc.	etc.
Scientific Notation	
$34,000 = 34 \times 10^3$	$0.043 = 4.3 \times 10^{-2}$
$4,925,000,000 = 4.925 \times 10^9$	$0.00000084 = 8.4 \times 10^{-7}$

Problems VII.
1. Find the squares of 4, 12, 17, 13.5, $\frac{1}{2}$, $\frac{2}{3}$.
2. Find the cubes of 3, 9, 18, 11.6, $\frac{3}{4}$.
3. By using Table I find the square root and cube root of 4, 8, 17, 83, and 56.
4. Perform the indicated operation:
   (a) $x^2 \cdot x^3$. (b) $(a^2x^3)(ayx^2)$.
   (c) $2^2 \div 2^3$. (d) $\dfrac{x^4}{x^2}$.
5. The floor space of a room is measured in square feet and is found by multiplying the length (measured in feet) times the width (measured in feet).
   (a) Find the floor space of the following rooms:
       (i) 9 ft. by 9 ft.   (ii) 12 ft. by 12 ft.
       (iii) 11.5 ft. by 11.5 ft.   (iv) x ft. by x ft.
   (b) A square room has the length equal to the width. Find the dimensions of the square rooms with the following floor space:
       (i) 64 sq. ft.   (ii) 169 sq. ft.   (iii) a^2 sq. ft.

The Formula and Equation

One of the most useful tools in mathematics is the concept of a formula. The reader may already be familiar with such formulas as: the area of a rectangle is the width times the length, which we express as $A = xy$; or: simple interest is given by $I = Prt$ (interest = principal × rate × time). Thus a **formula** *is an algebraic expression which states a fact, a law, a rule, etc.* In the formula we use the equal sign, =, and letters which represent specific quantities. Thus in the formula for the area of a rectangle, $A = xy$, A is the area, x is the length, and y is the width. The formulas are general expressions and are true for any value of the dimensions under discussion. That is, no matter what the length and the width of a rectangle are, we can always find the area by the formula $A = xy$. The letters x and y are then called *variables*. If we have measured a given rectangle, say a room, and found the width to be 8 feet and the length to be 12 feet, then we can find the area by letting $x = 8$ and $y = 12$ to give $A = xy = (8)(12)$ = 96 square feet. This is called an *evaluation* of a formula by *substituting* into the formula specific values for the variables.

Examples
(a) Given $A = \frac{1}{2}bh$; find A if $b = 10$ and $h = 3$.
   Solution.
   $$A = \tfrac{1}{2}bh = \tfrac{1}{2}(10)(3) = 15.$$

(b) Given $I = Prt$; find I if $P = 100, r = .03, t = 12$.

Solution.

$$I = Prt = (100)(.03)(12) = 36.$$

An **equation** *is a symbolic statement that two quantities are equal in value.*
It has two sides, a *left-hand* side and a *right-hand* side, which are
separated by an equal sign.

Examples

$$I = Prt.$$
$$x + 5 = 3x - 3.$$
$$3a + 4a = 7a.$$

If an equation is true for all values of the letters, it is called an
identity: e.g., $3a + 4a = 7a$. Otherwise it is called a **conditional
equation**; e.g., $x + 5 = 3x - 3$ is true only if $x = 4$, for then $4 + 5$
$= 12 - 3$ or $9 = 9$.

The statement that two quantities are equal in value is not changed
if

(a) **the same number is added to both sides;**
(b) **the same number is subtracted from both sides;**
(c) **both sides are multiplied by the same number;**
(d) **both sides are divided by the same number;**
(e) **both sides are raised to the same power.**

This is another way of saying that if a number is added to one side of
the equation, the same number must be added to the other side to
keep the equation the same. Similarly for subtraction, multiplication,
division, and raising to a power.

Examples

(a) Given $3 = 3$, then $3 + 2 = 3 + 2$ since $5 = 5$.
(b) Given $7 = 7$, then $7 - 4 = 7 - 4$ since $3 = 3$.
(c) Given $4 = 4$, then $3 \times 4 = 3 \times 4$ since $12 = 12$.
(d) Given $8 = 8$, then $8 \div 2 = 8 \div 2$ since $4 = 4$.
(e) Given $3 = 3$, then $3^2 = 3^2$ since $9 = 9$.
(f) Given $16 = 16$, then $\sqrt{16} = \sqrt{16}$ since $4 = 4$.
(g) Given $3x + 2 = 5$, then $3x + 2 - 2 = 5 - 2$ or $3x = 3$.

An important application of these rules is the changing of formulas.
The formula for the distance, d, that a car moves at a constant speed,
V, for a given time, t, is given by $d = Vt$. From this formula we can

find the distance if we are given V and t; e.g., if $V = 30$ miles per hour and $t = 2.5$ hours, then $d = (30)(2.5) = 75$ miles. If we divide both sides of this equation by V we have

$$\frac{d}{V} = t,$$

and now we can use this formula to find t if we know d and V. For example, how long would it take to go 200 miles at a constant speed of 50 miles per hour? We substitute into the formula to get

$$t = \frac{d}{V} = \frac{200}{50} = 4 \text{ hours.}$$

Example

The formula $F = \frac{9}{5}C + 32$ is used to change from centigrade degrees to Fahrenheit degrees in temperature measurements. If we subtract 32, multiply by 5, and then divide by 9 we have $C = \frac{1}{9}(5F - 160)$, which we use to change from Fahrenheit to centigrade.

The addition or subtraction of a number which appears in the equation leads to the useful process of *transposing*.

Example

Given the equation: $3x + a = 4b$.
Subtract a: $3x + a - a = 4b - a$
or $3x = 4b - a$.

We notice in this example that the term, a, which first appeared as $+a$ on the left side, now appears as $-a$ on the right side.

Rule. *To* **transpose** *a quantity from one side of the equation to the other side, simply remove it from the first side and write it on the other side* **with its sign changed.**

Example

Given: $6y - 3 = 4y + 7$.
Transpose $4y$: $6y - 4y - 3 = 7$.
Transpose -3: $6y - 4y = 7 + 3$.
Combine like terms: $2y = 10$.
Divide both sides by 2: $y = 5$.

The above example is a "sneak preview" of how to solve simple equations, which we shall discuss in the next section.

Problems VIII.

1. Given the formula $A = \sqrt{s(s-a)(s-b)(s-c)}$,
   where $s = \frac{1}{2}(a+b+c)$. Find A if $a = 3$, $b = 4$, and $c = 5$.
2. Given the formula $I = Prt$. Find I if $P = 500$, $r = .03$, $t = 5$.
   How much is $P + I$?
3. Given the formula $d = \frac{1}{2}gt^2$. Find d if $g = 32$ and $t = 2$. Find t
   if $d = 256$ and $g = 32$.
4. The distance traveled is equal to the average speed multiplied by
   the time, $d = vt$. Find the distance traveled by a car running at
   an average speed of 43.6 mi./hr. for 3.4 hours.
5. To change from Fahrenheit degrees to centigrade degrees we use
   the formula $C = \frac{5}{9}(F - 32)$. What is the temperature in centigrade if we measure 112 degrees Fahrenheit?
6. Find E if $E = \frac{1}{2}mv^2$ when $m = 180$ ibs. and $v = 25$ ft. per sec.
7. If two resistances, r_1, and r_2, are connected in parallel, the total
   resistance is found by $\frac{1}{R} = \frac{1}{r_1} + \frac{1}{r_2}$. Find R if $r_1 = 75$ ohms and
   $r_2 = 100$ ohms.
8. The cost for using electrical appliances can be expressed by the
   formula $C = wtr$, when w is watts; t, the time; and r, the rate.
   Find r if $C = 16$ cents, $w = 800$, and $t = 4$. With this value of r,
   how many hours can one burn a 50-watt light bulb for 25 cents?

Solution of Simple Equations

In a conditional equation the variable is often called the **unknown.**
Thus in the equation $x - 2 = 5$, the variable x is the unknown. To
solve the equation we try to find the value of x which makes the
equation an identity; e.g., if $x = 7$ in the equation $x - 2 = 5$,
then we have $7 - 2 = 5$ or $5 = 5$, which is an identity, and we say
$x = 7$ is a solution of this equation. The solution is also called the
root of the equation. Thus 7 is the root of the equation $x - 2 = 5$.
We shall also say that a root *satisfies the equation* and this means that
when we substitute the root into the equation we arrive at an identity.

Definition. *The **root** of an equation is that value of the unknown
which satisfies the equation.*

A simple equation in one unknown can be solved by applying the
rules we have already learned.

Examples

(a) Solve for x in the equation $x - 3 = 6$.

Solution.

Given: $x - 3 = 6.$
Add 3: $x - 3 + 3 = 6 + 3.$
Collect like terms: $x = 9.$

(b) Solve the equation $3x + 4 = 13$.

Solution.

Given: $3x + 4 = 13.$
Subtract 4: $3x = 13 - 4 = 9.$
Divide by 3: $x = 3.$

We shall now give a precise statement of what to do in solving the simple equation.

Procedure.

1. *Transpose all the terms containing the unknown to one side of the equation.*
2. *Transpose all the terms not containing the unknown to the other side of the equation.*
3. *Divide both sides of the equation by the coefficient of the unknown.*

Examples

(a) Solve $5x - 5 = 2x + 16$.

Solution.

Given: $5x - 5 = 2x + 16.$
Transpose: $5x - 2x = 16 + 5.$
Collect: $3x = 21.$
Divide: $x = 7.$

(b) Solve $15x + 17 = 8x - (3x + 2)$.

Solution.

Given: $15x + 17 = 8x - (3x + 2).$
Remove parentheses: $15x + 17 = 8x - 3x - 2.$
Transpose: $15x - 8x + 3x = -2 - 17.$
Collect: $10x = -19.$
Divide: $x = -1.9.$

In Example (b) above we could have combined $8x - 3x$ into $5x$ before we transposed. Such choice of sequences is left to the individual taste

and skill of the reader. The general advice is to do as many steps in
one operation as the person's skill permits and thus save writing and
time. Thus the solution of Example (b) could be written

$$15x + 17 = 8x - 3x - 2.$$
$$10x = -19,$$
$$x = -1.9.$$

Here we have removed the parentheses while writing the given equation; in the second step we transposed and collected the terms at the
same time. A word of caution: Don't combine so many steps that
mistakes are made. *It is more important to be right than to be fast.* We
shall present two examples without instructions; study each step to
have a complete understanding of what was done.

Examples

(a) Find x in $1.2x + .7a = .9x + 1.9a$.

   Solution.

$$1.2x + .7a = .9x + 1.9a.$$
$$12x - 9x = 19a - 7a.$$
$$3x = 12a.$$
$$x = 4a.$$

(b) Solve $2(3x + 4) = 3(x + 2) - 5.$

   Solution.

$$6x + 8 = 3x + 6 - 5.$$
$$3x = -7.$$
$$x = -\tfrac{7}{3}.$$

Problems IX.

1. Solve for x:
   (a) $2x - 4 = 8.$
   (b) $3x + 9 = 4x - 5.$
   (c) $x - 2(x + 5) = 3x - 5(x - 3).$
   (d) $1.3x - 2.2 = .7x + .8.$
   (e) $3a + 7x = 4x - 5a.$
   (f) $2(3x - 2) + a = 4x + 3(a - 4).$
2. Solve for C in the equation $F = \tfrac{9}{5}C + 32$ if $F = 212°$, $F = 32°$, and
   $F = 100°$.

Using Equations to Solve Problems

We learn how to solve an equation so that we may be able to solve
practical problems. However, we must also be able to translate from

the language of the problem into the language of mathematics. This usually means that we wish to write an equation from the statement of the problem and then to solve the equation.

Procedure.

1. *Read carefully the statement of the problem.*
2. *Pick out the unknown quantity and assign a letter to it.*
3. *If there is more than one unknown, find expressions which relate the unknowns.*
4. *Find two expressions that say the same thing and set them equal to each other.*
5. *Solve the equation obtained in step 4.*

Examples

(a) If 3 times a number is increased by 3, the result is 18. Find the number.

Solution.

Let N denote the unknown number.
Then $3N$ is 3 times the number.
The two equal expressions form the equation

$$3N + 3 = 18.$$

The solution is $N = 5$.

(b) Find three consecutive integers whose sum is 18.

Solution.

Let x be the first number; y, the second; z, the third. Consecutive integers are numbers which differ by one; e.g., 3, 4, and 5 are consecutive integers. Therefore $y = x + 1$ and $z = y + 1 = x + 2$. The equality statement of the problem says

$$x + y + z = 18$$

or

$$x + (x + 1) + (x + 2) = 18$$

so that

$$3x + 3 = 18$$

or

$$x = 5.$$

Check: $5 + 6 + 7 = 18.$

(c) A sum of money amounting to $4.30 consists of nickels, dimes, and quarters. There are four times as many dimes as nickels and six fewer quarters than dimes. How many of each coin are there?

Solution.

Let x be the number of nickels, then $4x$ is the number of dimes and $4x - 6$ is the number of quarters. The equality statement is

$$.05x + .10(4x) + .25(4x - 6) = 4.30$$

since the value of a nickel is $0.05; a dime, $0.10; etc.

We solve this equation as follows:

$$5x + 40x + 100x - 150 = 430.$$
$$145x = 580.$$
$$x = 4.$$

Thus we have 4 nickels, $4x = 16$ dimes, and $4x - 6 = 10$ quarters.

We notice in the above examples that often an additional fact must be known. Thus in example (b) we had to know what consecutive integers were; in example (c) we had to know the worth of nickels, dimes, and quarters. Such facts come from the general knowledge and experience of the reader. A very useful fact is:

The whole of any quantity equals the sum of its parts.

Example

A feed merchant wants to obtain 200 bushels of a mixture of wheat worth $1.00 per bushel by mixing wheat worth $1.35 per bushel and $0.75 per bushel. How much of each kind should be used?

Solution.

Let x be the number of bushels of $1.35 wheat, then $200 - x$ is the number of bushels of $0.75 wheat. The total worth is the sum of the worth of the individual parts so that

$$1.35x + .75(200 - x) = 200(1.00).$$
$$135x + 15000 - 75x = 20000.$$
$$60x = 5000.$$
$$x = 83\tfrac{1}{3}.$$

Therefore $83\tfrac{1}{3}$ bushels of $1.35 wheat and $200 - 83\tfrac{1}{3} = 116\tfrac{2}{3}$ bushels of $0.75 wheat is needed.

Problems X.

1. Find the cost per book if six books at x dollars per book cost $16.50.

2. If 6 is added to $5x$ and 12 subtracted from the sum, the result equals $3x$ plus 4. Find x.

3. A car runs 17.5 miles on one gallon of gasoline costing 31.9¢ per gallon. How much will be spent on gasoline to go 385 miles?

4. Find three consecutive even integers whose sum is 132. (Hint: If x is the first even integer, then $x + 2$ is the next even integer.)

5. A sum of money amounting to $6.45 consists of nickels, dimes, quarters, and half dollars. There are half as many quarters as nickels, 2 more dimes than nickels, and two more half-dollars than quarters. How many of each coin are there?

6. Two cars 240 miles apart are driven toward each other. One car averages 30 miles an hour and the second car averages 50 miles an hour. How long before they meet? (Hint: distance = average speed × time.)

7. Mary saves twice as much money a week as Tom, who saves three fifths of his earnings. How long will it take Mary to save $54 if Tom earns $15 a week?

8. The total cost of a dress and a coat is $98. If the dress cost $22 less than the coat, how much did each cost?

9. The sum of a number plus its double plus its triple is 36; what is the number?

10. Two angles are said to be complementary if their sum is 90°. Find the complementary angles whose difference is 14 degrees.

11. If a man sold 2 acres more than $\frac{2}{3}$ of his land and had 4 acres less than $\frac{1}{2}$ of it left, how many acres did he start with?

12. If sirloin steak costs $1\frac{1}{2}$ times as much as round steak and the total cost of 2 pounds of sirloin and 3 pounds of round was $4,74, what was the cost of each per pound?

13. A man bought a herd of cattle for $12,000 and then sold all but 20 of them at $40 per head. If he got his $12,000 back for those he sold, how many did he buy?

14. A chemist has 83.4 cubic centimeters (cc.) of hydrocloric acid. If he desires to separate the liquid into two parts such that the larger amount exceeds the smaller by 17.2 cc., how many cc. must each part contain?

15. If a board 12 feet long is to be cut into two pieces such that one is $2\frac{2}{3}$ times as long as the other, what is the length of each piece?

16. If a man spent $\frac{1}{4}$ of his money on railroad fares, $\frac{1}{2}$ of it for lodging,

$48 for all other expenses and returned home with $5, how much did he start with?

17. A man gave his son 25 cents for every dime his son saved. At the end of a year the boy had $70.00. How much did his father contribute?

18. A boy scout troup of 30 boys is to be divided into two groups such that one group has 6 more than the other. How many boys are there in each group?

19. A father employed his son for 30 days, paying him $4.50 a day for each day worked but collecting a forfeit of $2.00 a day for each day absent. If the boy was paid $70 how many days did he work and how many days was he absent?

Algebra of Polynomials

We have already learned that an algebraic expression containing two or more terms is called a *polynomial*. We have also learned that in adding and subtracting algebraic terms we can combine only *like* terms. Let us now study multiplication of algebraic expressions.

I. To multiply a monomial by a monomial:

   (i) *Determine the sign of the product.*

   (ii) *Multiply the numerical coefficients.*

   (iii) *Multiply like letters by adding their exponents.*

Examples

   (a) $(12a^2b^4)(2ab^3) = 24a^3b^7$.

   (b) $(9xy^3)(-4x^2y) = -36x^3y^4$.

   (c) $(-17ax^2y)(-2a^2xy) = 34a^3x^3y^2$.

   (d) $(-3ax)(6ay) = -18a^2xy$.

II. To multiply a polynomial by a polynomial:

   (i) *Multiply every term of the multiplicand by each term of the multiplier.*

   (ii) *Write the like terms of the partial products under each other.*

   (iii) *Find the algebraic sum of the partial products.*

Examples

   (a) Multiply $x + 5$ by $x - 6$.

     Solution.

$$x + 5$$
$$\underline{x - 6}$$

x times $(x + 5)$: $x^2 + 5x$

-6 times $(x + 5)$: $\underline{-6x - 30}$

Add: $x^2 - x - 30$. *Ans.*

(b) Multiply $x^2 - 2xy + 3y^2$ by $3x^2 - 2xy$.

Solution.

$$x^2 - 2xy + 3y^2$$
$$3x^2 - 2xy$$

Multiplicand $(3x^2)$: $\overline{3x^4 - 6x^3y + 9x^2y^2}$

$(-2xy)$: $\underline{\quad - 2x^3y + 4x^2y^2 - 6xy^3}$

$3x^4 - 8x^3y + 13x^2y^2 - 6xy^3$. *Ans.*

Let us now turn to the division of algebraic expressions.

I. To divide a monomial by another:

 (i) *Determine the sign of the quotient.*

 (ii) *Divide the numerical coefficient of the dividend by that of the divisor.*

 (iii) *Determine the exponents of the letters.*

Examples

(a) Divide $30a^5x^3$ by $-5a^2x$.

Solution.

A plus divided by a minus yields a minus,
30 divided by 5 is 6,
a^5 divided by a^2 is $a^{5-2} = a^3$,
x^3 divided by x is $x^{3-1} = x^2$.
Therefore $30a^5x^3 \div (-5a^2x) = -6a^3x^2$. *Ans.*

(b) Divide $65xy^6z^7$ by $13xyz^3$.

Solution.

$$\frac{65xy^6z^7}{13xyz^3} = 5y^5z^4. \quad Ans.$$

II. To divide one polynomial by another:

 (i) *Arrange each polynomial in descending powers of a common letter.*

 (ii) *Divide the first term of the dividend by the first term of the divisor to obtain the first term of the quotient.*

 (iii) *Multiply the entire divisor by the first term of the quotient and subtract the product from the dividend.*

 (iv) *The remainder is a new dividend and steps (ii) and (iii) are repeated on this dividend.*

 (v) *Continue until a remainder is obtained which is an expression whose first term does not contain the first term of the divisor as a factor.*

(vi) *Check the result by replacing the letters by numbers; do not use numbers which make the divisor zero.*

Examples

(a) Divide $14x^3 + 17x^2 - 23x - 3$ by $2x + 5$.

Solution.

$$
\begin{array}{r|l}
\textbf{(Dividend)} \quad 14x^3 + 17x^2 - 23x - 3 & \; 2x + 5 \quad \textbf{(Divisor)} \\
\underline{14x^3 + 35x^2 \qquad\qquad} & \; 7x^2 - 9x + 11 \quad \textbf{(Quotient)} \\
 -18x^2 - 23x - 3 & \\
 \underline{-18x^2 - 45x \qquad} & \\
 22x - 3 & \\
 \underline{22x + 55} & \\
 - 58 \quad \textbf{(Remainder)} &
\end{array}
$$

Check: Let $x = 1$.

Dividend $= 14(1) + 17(1) - 23(1) - 3 = 5$.
Divisor $= 2(1) + 5 = 7$.
Quotient $= 7(1) - 9(1) + 11 = 9$.
Remainder $= -58$.
Dividend $=$ Divisor $\times$ Quotient $+$ Remainder.
$$5 = (7)(9) - 58 = 63 - 58 = 5.$$

(b) Divide $15x^3 + 23x - 26x^2 - 3$ by $3x^2 + 3 - 4x$.

Solution.

$$
\begin{array}{r|l}
15x^3 - 26x^2 + 23x - 3 & \; 3x^2 - 4x + 3 \\
\underline{15x^3 - 20x^2 + 15x \quad} & \; 5x - 2 \quad \textbf{(Quotient)} \\
 -6x^2 + 8x - 3 & \\
 \underline{-6x^2 + 8x - 6} & \\
 3 \quad \textbf{(Remainder)} &
\end{array}
$$

The check is left to the reader.

Problems XI.

1. Multiply

    (a) $14a^2x$ by $3ax^2$.   (b) $3x^2$ by $13axy$.
    (c) $7axy^2$ by $3abx^3$.   (d) $21xy$ by $-ay$.
    (e) $3x^2 - 2$ by $2x - 3$.   (f) $2x + 1$ by $2x + 3y$.
    (g) $x - 3$ by $x + 3$.   (h) $2x - 3y$ by $2x + 3y$.
    (i) $3x^3 - 4x^2y + 2xy^2 - y^3$ by $2x + 3y$.

2. Divide

    (a) $18a^2x$ by $2ax$.   (b) $9ax^5y^3$ by $3ax^3y^2$.
    (c) $27a^3x^5y^4$ by $-9a^2x^3y^4$.   (d) $125x^4y^3$ by $-25x^2y^2$.
    (e) $4x^4 - 19x^3y + 2x^2y^2 + xy^3 - 6y^4$ by $-4xy - 3y^2 + x^2$.

Factoring

The process of finding two or more expressions whose product is a given expression is called **factoring**. This is accomplished by being able to recognize a group of typical products, and upon seeing them in an algebraic expression we can immediately write down the factors. The first step is to memorize a few products. We shall list some common ones:

1. $ab + ac = a(b + c)$.
2. $a^2 + 2ab + b^2 = (a + b)^2$.
3. $a^2 - 2ab + b^2 = (a - b)^2$.
4. $a^2 - b^2 = (a - b)(a + b)$.
5. $x^2 + (a + b)x + ab = (x + a)(x + b)$.
6. $acx^2 + (ad + bc)x + bd = (ax + b)(cx + d)$.
7. $a^3 + b^3 = (a + b)(a^2 - ab + b^2)$.
8. $(a^3 - b^3) = (a - b)(a^2 + ab + b^2)$.

To ease the process of factoring, first remove all common expressions.

Examples

(a) Find the factors of $3ax^2 - 27ay^2$.

Solution.

Remove $3a$: $3ax^2 - 27ay^2 = 3a(x^2 - 9y^2)$.
Since $9y^2 = (3y)^2$ we use (4): $= 3a(x - 3y)(x + 3y)$. *Ans.*

(b) Find the factors of $x^2 + x - 2$.

Solution.

By (5): $x^2 + x - 2 = (x - 1)(x + 2)$.

(c) Find the factors of $2x^2 + 5x - 12$

Solution.

The factors of 2 are 1 and 2.
The factors of 12 are $(1, 12)$ or $(2, 6)$ or $(3, 4)$ and since we have -12 one of the two factors must be negative. We now search for combinations such that the sum of the products of the factors of 2 and -12 yields 5. We find

$$2x^2 + 5x - 12 = (2x \pm ?)(x \pm ?)$$
$$= (2x - 3)(x + 4).$$

Problems XII.

1. Factor the following expressions:
   (a) $x^2 - y^2$.
   (b) $3ax + 6axy$.
   (c) $x^2 - x - 6$.
   (d) $3ax^2 + 6ax + 3a$.

(e) $3x^3 + 3y^3$.   (f) $x^4 - 81y^4$.

(g) $2x^2 - 5x - 12$.   (h) $18x^2 + 45x - 108$.

The Quadratic Equation

We have already discussed the solution of equations in which the unknown appears to the first power. In this section we shall consider an equation in which the unknown appears to the second power; such an equation is called a **quadratic equation.** The general form is

$$ax^2 + bx + c = 0,$$

where a, b, and c are constants and x is the unknown.

Before we study the solution of a quadratic equation let us consider the statement.

If the product of two factors equals zero, then either or both factors equals zero.

The truth of the statement can be realized by considering the expression

$$(a)(b) = 0.$$

Now if $a = 0$,   we have $(0)(b) = 0$,

if $b = 0$,   we have $(a)(0) = 0$,

if $a = 0$ and $b = 0$,   we have $(0)(0) = 0$,

all of which are true statements by our definition for multiplication by zero.

We shall consider two ways to solve the quadratic equation.

I. By Factoring. To apply this method we factor the quadratic expression which is equal to zero, then set each factor equal to zero and solve the first order equation.

Example

Solve $x^2 - x - 6 = 0.$

Solution.

Given: $x^2 - x - 6 = 0.$

Factor: $(x - 3)(x + 2) = 0.$

Set each factor $x - 3 = 0.$

equal to zero: $x + 2 = 0.$

Solve: $x = 3$ and $x = -2$. *Ans.*

We notice that there are two answers to such problems, which can be proved by the fundamental theorem of algebra. However, we shall not prove it in this book, but shall simply accept the fact that every **quadratic equation has two roots.**

The method of factoring is used only if we can easily factor the quadratic expression. It is also important to note that the equation must be written in the form which sets the right-hand side equal to zero.

Example

Solve $2x^2 = 5x + 12$.

Solution.

Transpose: $2x^2 - 5x - 12 = 0$.
Factor: $(2x + 3)(x - 4) = 0$.
Solve: $2x + 3 = 0$ or $x = -\frac{3}{2}$.
            $x - 4 = 0$ or $x = 4$.

II. By the Quadratic Formula. The solution of the general quadratic equation

$$ax^2 + bx + c = 0$$

is given by

$$x = \frac{-b \pm \sqrt{b^2 - 4ac}}{2a}.$$

Notice that there are two roots, one with the $+$ sign before the radical and one with the negative sign. We can now use this formula to find the roots by simply identifying the a, b, and c and substituting their values into the formula.

Examples

(a) Solve $2x^2 - 6x + 3 = 0$.

Solution.

Here $a = 2$, $b = -6$, and $c = 3$.

Substituting: $x = \dfrac{-(-6) \pm \sqrt{36 - 4(2)(3)}}{2(2)}$

$$= \frac{6 \pm \sqrt{36 - 24}}{4}$$

$$= \frac{6 \pm \sqrt{12}}{4} = \frac{6 \pm 2\sqrt{3}}{4}$$

$$= \frac{2(3 \pm \sqrt{3})}{4} = \frac{3 \pm \sqrt{3}}{2}.$$

(b) Solve $3x^2 = 7x - 2$.

Solution.

Rewrite: $3x^2 - 7x + 2 = 0$.

Then $a = 3, b = -7, c = 2$.

Substituting: $x = \dfrac{7 \pm \sqrt{49 - 24}}{6}$

$$= \dfrac{7 \pm \sqrt{25}}{6} = \dfrac{7 \pm 5}{6}$$

Separating the roots: $x = \dfrac{7 + 5}{6} = \dfrac{12}{6} = 2$.

$$x = \dfrac{7 - 5}{6} = \dfrac{2}{6} = \dfrac{1}{3}.$$

(c) Solve $10x^2 = 13x - 3$.

Solution.

Rewrite: $10x^2 - 13x + 3 = 0$.

Substitute: $x = \dfrac{13 \pm \sqrt{169 - 120}}{20}$

$$= \dfrac{13 \pm \sqrt{49}}{20} = \dfrac{13 \pm 7}{20}.$$

Separate roots: $x = \dfrac{13 + 7}{20} = \dfrac{20}{20} = 1$.

$$x = \dfrac{13 - 7}{20} = \dfrac{6}{20} = \dfrac{3}{10}.$$

If $b = 0$ the x term is missing from the quadratic equation and we have a pure quadratic equation. The solutions of this equation can be obtained by transposing the constant term to the right side, dividing by a, and taking the square root of both sides.

Example

Solve $4x^2 - 64 = 0$.

Solution.

Rewrite: $4x^2 = 64$.

Divide by 4: $x^2 = 16$.

Take square root: $x = \pm 4$.

In solving the quadratic equation by the formula, the numbers a, b, and c could be such that $b^2 - 4ac$ would be a negative number. We would then be required to find the square root of a negative number, but this we cannot do with the number system that we have defined so far. For example, neither $(-2)^2$ nor $(+2)^2$ would give us -4, so what is $\sqrt{-4}$? To answer this question we must extend our number system. We shall call $\sqrt{-1}$ an **imaginary** number and *define* the letter i to be a number such that

$$i^2 = -1.$$

In other words $\sqrt{-1} = i$. Now we can write $\sqrt{-4} = \sqrt{4(-1)}$ $= \sqrt{4}\sqrt{-1} = 2i$. This imaginary number has some interesting properties.

$$i = \sqrt{-1}.$$
$$i^2 = -1.$$
$$i^3 = i^2 i = -i.$$
$$i^4 = i^2 i^2 = (-1)(-1) = 1.$$
$$i^5 = i^4 i = (1)i = i.$$
$$i^6 = i^4 i^2 = (1)(-1) = -1.$$
$$\text{etc.}$$

If we combine an imaginary number with a real number we have a **complex number,**

$$a + bi,$$

where a is called the *real part* and bi is called the *imaginary part*.

We now have a number system which will contain all the roots of a quadratic equation.

Example

Solve $x^2 - 4x + 13 = 0$.

Solution.

$$a = 1, b = -4, c = 13.$$

Substitute: $x = \dfrac{4 \pm \sqrt{16 - 52}}{2} = \dfrac{4 \pm \sqrt{-36}}{2}$

$$= \frac{4 \pm 6i}{2} = 2 \pm 3i. \ Ans.$$

Problems XIII.

Solve the following quadratic equations.

1. $3x^2 - 75 = 0$.
2. $x^2 - 7x + 12 = 0$.
3. $3x^2 - 21x + 30 = 0$.
4. $6x^2 - 13x + 6 = 0$.
5. $3x^2 = 15 - 4x$.
6. $x^2 + x + 1 = 0$.
7. $6x^2 + x - 2 = 0$.
8. $x^2 - 4x + 1 = 0$.
9. $36x^2 - 36x + 13 = 0$.
10. $8x^2 + 26x + 21 = 0$.
11. $9x^2 + 12x + 4 = 0$.
12. $4x^2 - 4x - 11 = 0$.
13. $16x^2 - 24x + 13 = 0$.
14. $x^2 + 2x + 3 = 0$.

15. The square of a number plus 12 is seven times the number. Find the numbers.

16. The formula $s = v_0t + \frac{1}{2}gt^2$ gives the distance traveled by a falling body having an initial velocity of v_0 feet per second; t is the time in seconds. If $v_0 = 280$ ft./sec. and g (the force of gravity) $= 32$ how long will it take a body to fall 3000 feet?

17. The impedance Z of an alternating current circuit is given by $Z^2 = R^2 + X^2$ where R is the resistance of the circuit and X the reactance. Find R if $Z = 50$ ohms and $X = 40$ ohms.

18. The formula $y(4v^2 + 5v - 2) = 1200\,Hd$ is used to find the velocity in feet per second of the flow of water from a pipe lead. Find v if $y = 100$, $d = \frac{1}{2}$, and $H = 4$.

19. The area of a rectangle is the length times the width, $A = ab$. Find the dimensions of a rectangular picture whose area is 18 square inches if the length is 3 inches more than the width.

20. Find a number such that the sum of the number and its reciprocal (one over the number) is $4\frac{1}{20}$.

Ratio and Proportion

It is often necessary to compare two quantities. One method of comparison is to say that one quantity is $\frac{2}{3}$ as long as a second, or a house is one and a half times as big as another. This method of comparison is the ratio method. Thus a ratio is an expression used to compare two quantities of the same kind. A **ratio** is simply a fraction; that is, if the ratio of the length of one room to that of another is 4 to 5, then one room is $\frac{4}{5}$ as long as the other. In our language of mathematics we often write a ratio in the form 4:5 which is read "4 is to 5" and here 4 is the numerator and 5 is the denominator. We stress again that a ratio compares like quantities; thus we can use a ratio to compare inches to inches but not to compare a length measured in inches with a length measured in feet unless we change to the same dimensions.

Examples

(a) A room measures 12' by 15'. What is the ratio of its dimensions?

Solution. The ratio is $\frac{12}{15} = \frac{4}{5}$. Thus the width is $\frac{4}{5}$ the length.

(b) An advertisement shows a picture of an article which is labeled "$\frac{3}{5}$ actual size." If the length on the picture measures $2\frac{1}{2}$ inches, what is the actual length of the article?

Solution. Change $2\frac{1}{2}$ to $\frac{5}{2}$ and write the ratio

$$\tfrac{5}{2}:\tfrac{3}{5} = \left(\tfrac{5}{2}\right)\left(\tfrac{5}{3}\right) = \tfrac{25}{6}.$$

Thus the article is $4\frac{1}{6}$ inches long.

Sometimes it is possible to compare *four quantities* by comparing them in *pairs* as ratio. *A* **proportion** *is the equality of two or more ratios.* Thus $2:5 = 6:15$ is a proportion. The symbols which are frequently used are

:, the symbol for "is to"
::, the symbol for "as."

Thus

$$2:5::6:15$$

is read

"2 is to 5 as 6 is to 15."

In fractional form this proportion is written

$$\tfrac{2}{5} = \tfrac{6}{15}$$

and we see that : is the same as the division sign and :: is the same as the equal sign.

Let us consider the general proportion

$$a:b::c:d.$$

The "end" terms of the proportion are called **extremes**. The "middle" terms of the proportion are called **means**.

Property. *In a proportion, the product of the means is equal to the product of the extremes;*

$$bc = ad.$$

This is easily seen if we change the ratios to fractions and write

$$\frac{a}{b} = \frac{c}{d}$$

and multiply both sides of this equation by bd.

$$(bd)\left(\frac{a}{b}\right) = (bd)\left(\frac{c}{d}\right)$$

and after cancelling we have

$$ad = bc.$$

If one of the four quantities is unknown, we can now obtain a simple equation from which we can solve for that unknown.

Algebra

Examples

(a) If 12 tons of coal cost \$72, what will 27 tons cost at the same rate per ton?

Solution. Let x stand for the number of dollars that 27 tons will cost; then the ratio of the numbers of tons is equal to the ratio of costs. We can then write

$$12:27 = 72:x$$

and

$$12x = (27)(72)$$

or

$$x = \frac{(27)(72)}{12} = \$162.00.$$

(b) A road map shows a scale of $1'' = 50$ mi. The distance between two cities was measured to be $3\frac{1}{4}$ inches. How far apart are the cities?

Solution.

The proportion is $3\frac{1}{4}:1 = x:50$.
By the property: $x = 50(3\frac{1}{4}) = 162.5$ mi.
Note that the dimensions of each ratio are the same; that is, $3\frac{1}{4}:1$ is in inches and $x:50$ is in miles if x is in miles.

(c) If 15 men can do a job in 28 days, in how many days can 21 men do the same job?

Solution. Since more men can do the job in less time, the ratio of the number of men is equal to the *inverse* ratio of the number of days. Let x be the number of days for 21 men, then the proportion is

$$15:21 = x:28$$

in which we have inverted the second ratio. Thus we have

$$21x = (15)(28)$$
$$x = \frac{(15)(28)}{21} = 20.$$

Therefore 21 men can do the job in 20 days.

Problems XIV.

1. Find the unknown term in the following proportions:
   (a) $14:22 = 7:x$. (b) $33:12 = x:28$.
   (c) $2.5:x = 0.5:3$. (d) $x:11 = 30:55$.

2. If 7 inches is to $\frac{1}{2}$ foot as x is to $1.25, what is x? (Note the dimensions.)

3. A baseball player makes 4 hits in 12 times at bat. Express this as a ratio and then change the ratio to a decimal fraction.

4. If a man who earns $6400 a year pays $1200 per year for rent, what is the ratio of his rent to his income?

5. A farmer bought 900 chicks from one hatchery and raised 795 of them. He also bought 500 chicks from another hatchery and raised 465 of these. On the basis of ratios which hatchery appears to have the healthier chicks?

6. The efficiency of a machine is the ratio of the output to the input. What is the efficiency of an engine if the input is 96 horse-power and the output is 88 horsepower? (Express to two decimals.)

7. If an automobile engine is .27 efficient, what is the input in horse-power to get an output of 150 horsepower?

8. If 5 men can do a job in 30 days, how long will it take 25 men to do it?

9. If it takes 16 gallons to drive a car 288 miles, how many gallons will it take to drive 558 miles at the same rate?

10. If the mixture of an alloy is 2 parts of copper, 3 parts of lead, and 4 parts of tin, how many pounds of each are there in a mixture weighing 102 pounds?

11. If a boat drifts downstream 20 miles in 8 hours, how far will it drift in 12 hours?

Review Exercises II.

1. Remove symbols of grouping and simplify.
   (a) $2x - \{2 + [3x - 2(x + 7)] - x\}$.
   (b) $6[3 + 2a - \{2a - 3(a - 1)\} + 3a]$.
   (c) $(3y + 2) - [2y + 6(y - 3)]$.

2. Perform the indicated operation.
   (a) $(3^2)(3^4) - (2^3)(2^5)$.
   (b) $(2x^2)^3(x^2) - 4x^3$.
   (c) $(2x^{-4})(x^3)$.

3. By using Table I, find the square root and cube root of 16, 27, 98, and 7.

4. Evaluate the following formulas:
   (a) $d = 16t^2 + V_0 t + d_0$, if $V_0 = 36$, $d_0 = 200$, and $t = 3$.
   (b) $P = br$, if $b = 1200$ and $r = .04$.
   (c) $I = Prt$, if $P = 2000$, $r = .035$, and $t = 5$.

5. Solve the following equations:
   (a) $3x - 5 = x + 7$.
   (b) $2(6x - 5) + 3 = 8x - 5$.
   (c) $x - 3 = 4x$.

6. Perform the indicated operations:
   (a) $(3ax^2 - y)(2ax - 3y)$.
   (b) $(6ax - 2a^2x^2 - 3a^3x^3)(2 + x)$.
   (c) $(14a^3x^3 - 7a^2x^2 + 9ax + 9) \div (2ax + 1)$.

7. Factor the following expressions:
   (a) $4x^2 - 16y^2$.
   (b) $6x^2 - x - 35$.
   (c) $3ax^2 - ax - 4a$.

8. Solve the following equations:
   (a) $6x^2 - x - 35 = 0$.
   (b) $x^2 + 6x + 7 = 0$.
   (c) $4x - 5 = x^2$.

9. The product of two consecutive even numbers is 48; find the numbers.

10. Find the ratio of the number of hits to the times at bat to 3 places for each of the following players:

Player	At Bat	Hits	Player	At Bat	Hits
A	136	48	G	140	40
B	120	36	H	144	36
C	152	44	I	98	28
D	115	30	J	110	20
E	148	44	K	160	40
F	150	50	L	40	12

Chapter III

PERCENTAGE

Per Cent

In its wider sense percentage is the study of per cent and its uses. It is in common usage in banking, business, sales, taxes, and many other everyday practices. Let us begin the study by defining our terms.

The term **per cent** *is another way of saying* **hundredths.** Per cent is expressed by the sign: %.

We now have three forms of expressing hundredths:

$\frac{1}{100}$, the common fraction form;
.01, the decimal fraction form;
1%, the per cent form;

which are all equivalent. Since per cent means hundredths, it is used like hundredths. To change per cent to a decimal fraction form, drop the per cent sign and place a decimal point two places to the left.

Examples

$$7\% = 0.07; \ 37\tfrac{1}{2}\% = 0.375;$$
$$19\% = 0.19; \ 125\% = 1.25.$$

To change a decimal fraction to a per cent, place the decimal point two places to the right and annex the per cent sign.

Examples

$$0.09 = \ \ 9\%; \ 0.667 = 66.7\%;$$
$$0.42 = 42\%; \ 2.39 \ = 239\%$$

In our terminology per cent is usually used in a phrase; e.g., *"5 per cent of $100 is $5."* In these cases we *change per cent to hundredths and multiply.*

Examples

(a) How much is 12% of $450?

Solution: ($450)(.12) = $54.

(b) If $37\frac{1}{2}$% of the total weight of 46,312 pounds for a two-stage rocket is in the second stage, how heavy is the second stage?

Solution: (46,312)(.375) = 17,367 pounds.

HANDY ANDY No. 8		
Per Cents as Fractions		
$5\%\ =0.05\ =\frac{1}{20}$	$33\frac{1}{3}\%=0.33\frac{1}{3}=\frac{1}{3}$	$60\%\ =0.60\ =\frac{3}{5}$
$8\frac{1}{3}\%=0.08\frac{1}{3}=\frac{1}{12}$	$37\frac{1}{2}\%=0.375=\frac{3}{8}$	$62\frac{1}{2}\%=0.625=\frac{5}{8}$
$12\frac{1}{2}\%=0.125=\frac{1}{8}$	$40\%\ =0.40\ =\frac{2}{5}$	$75\%\ =0.75\ =\frac{3}{4}$
$16\frac{2}{3}\%=0.16\frac{2}{3}=\frac{1}{6}$	$41\frac{2}{3}\%=0.41\frac{2}{3}=\frac{5}{12}$	$80\%\ =0.80\ =\frac{4}{5}$
$20\%\ =0.20\ =\frac{1}{5}$	$50\%\ =0.50\ =\frac{1}{2}$	$83\frac{1}{3}\%=0.83\frac{1}{3}=\frac{5}{6}$
$25\%\ =0.25\ =\frac{1}{4}$	$58\frac{1}{3}\%=0.58\frac{1}{3}=\frac{7}{12}$	$87\frac{1}{2}\%=0.875=\frac{7}{8}$

Percentage

Let us be more rigorous in our discussion of percentage. We begin by defining three basic terms.

*The **rate** is the given per cent. It is a ratio expressed in hundredths. The **base** is the number with which a comparison is made. In a discussion of percentages it is the number of which a certain per cent is found. The **percentage** is the result obtained when the base is multiplied by the rate expressed in hundredths.*

We can now express the basic formula

$$p = rb$$

where

p is the percentage,
r is the rate,
b is the base.

Given any two of the quantities we can find the third by solving the resulting equation.

I. *To find the percentage.* If Tom sold 85% of his 40 papers, how many papers did he sell?

Given: $r = 85\% = 0.85$ and $p = 40$.
Then $p = 40(0.85) = 34.00$ or Tom sold 34 papers.

II. *To find the rate.* Mac's Food Mart sold $198,732 worth of merchandise and made a profit of $24,841.50. What per cent of the sales was the profit?

Given: $b = 198,732$ and $p = 24,841.50$.
Then $24841.50 = 198732r$
and $r = .125$ or $12\frac{1}{2}\%$.

III. *To find the base.* A salesman spends 7% of his earnings on his car. If his car expense was $64.75 what were his earnings?

Given: $r = 7\% = .07$ and $p = 64.75$.
Then $64.75 = .07b$ or $b = 925$.
His earnings were $925.00.

The batting average of baseball players is calculated by dividing the number of hits by the number of times at bat correct to 3 places.

Examples

(a) The leading batter of the Detroit Tigers has 137 hits in 386 times at bat. What is his batting average?

Solution. Let $b = 386$ and $p = 137$, then
$$137 = 386r, \text{ or } r = .3549.$$

Rounding off we have a batting average of .355.

(b) The leading batter of the National League has been at bat 415 times and has a batting average of .400. How many hits has he made?

Solution. Let $b = 415$ and $r = .400$, then
$$p = 415(.400) = 166.$$

He has made 166 hits.

Increase and Decrease

The per cent that some quantity has become greater is called the **rate of increase** and the amount by which it has increased is called the **amount of increase**. Similarly, if a quantity has become smaller we have the **per cent of decrease** and the **amount of decrease**. The formula

$$p = rb$$

can be used if we let p be the amount of increase or decrease, b be the quantity that was changed, and r be the per cent of increase or decrease.

Examples

(a) The rainfall for the month of May last year was 12.3 inches and this year was 11.4 inches. What was the amount and rate of decrease from last year's rainfall?

Solution.

Amount of decrease $12.3 - 11.4 = 0.9$ in.
Since we are comparing with last year's rainfall, $b = 12.3$ and $0.9 = 12.3r$.
Then $r = .073$ or about 7%.

(b) The cost of a certain brand of shoes has increased 8.6%. If the cost was $8.88 a pair what is the amount of increase and the new cost?

Solution.

Given: $b = 8.88$ and $r = 0.086$.
Then amount of increase $= p = 8.88(0.086) = 0.76368$.
This is rounded off to $0.76.
The new price is $8.88 + $0.76 = $9.64.

(c) A car costing $2875 could be bought for $800 down and $99.00 a month for 24 months. What is the amount and rate of increase if the car was bought on the installment plan?

Solution.

The cost of the car on the installment plan is
$$(\$99)(24) + \$800 = \$3176.$$
The amount of the increase is
$$\$3176 - \$2875 = \$301.$$
The rate of increase is found by
$$301 = 2875r \quad \text{or} \quad r = .1047$$
so that the rate of increase is about $10\frac{1}{2}$%.

Problems XV.

1. Change each of the following to a decimal fraction:
   (a) 12%.   (b) $67\frac{1}{2}$%.   (c) 4.5%.   (d) 0.5%.
2. Change each of the following to a per cent:
   (a) 0.15.   (b) 1.25.   (c) 0.004.   (d) $\frac{5}{8}$
3. Find the following amounts:
   (a) $12\frac{1}{2}$% of 932.   (b) 150% of 50.
   (c) $33\frac{1}{3}$% of 273.   (d) 4.5% of 28.375.

4. Mr. Jones pays $75.00 per month for rent. What is his salary if this amount is 15% of it?

5. A salesman spends 8% of his monthly salary on his car. If his salary is $900 a month, how much does he spend on his car?

6. Mrs. Smith spends $36.75 a week on groceries. If Mr. Smith earns $180.25 a week, what per cent of his earnings is spent on groceries?

7. A suit which ordinarily sells for $60.00 was put on sale for $49.75. What was the rate of decrease?

8. The rate of traffic on 38th Street was increased 15% of its regular flow of 300 cars per hour. What was the amount of increase and what is the new flow of cars per hour?

9. A bicycle costs $57.00 if purchased for cash, but on the installment plan the same bicycle costs $69.00. What is the rate of increase in the price if bought on the installment plan?

10. This week's sales totaled $3361.77 and last week's were $3112.75. Find the amount and rate of increase.

Discounts

The amount that can be *taken off* the original price is called a **discount**. Stores will frequently advertise sales, "Prices reduced 20%" or "Special Sale — $33\frac{1}{3}\%$ off." Discounts are also given to buyers for paying cash or paying before a certain date. Discount is calculated in the same way as percentage.

Examples

(a) A Fire Sale advertised 25% off. If you bought two articles marked $9.00 and $15.00, how much should you pay?

Solution.

The total list price is $9.00 + $15.00 = $24.00.
The discount is 25%. Therefore

$$24.00 \times .25 = 24.00(\tfrac{1}{4}) = \$6.00$$

is the amount of the discount. The cost is

$$\$24.00 - \$6.00 = \$18.00. \quad Ans.$$

(b) A merchant's bill reads "8% discount if paid before the 10th of the month." If he purchased $1230.00 worth of goods, how much should he pay if he pays the bill on the 5th of the month?

Solution.

$$\$1230 \times .08 = \$98.40.$$
$$\$1230.00 - \$98.40 = \$1131.60. \quad Ans.$$

Buyers from manufacturers are often given more than one discount which are listed in sequence; for example, "40% and 10% off." This does not mean 50% off but means that first 40% is deducted from the list price and then 10% of that result is deducted. These discounts are called *chain discount, series discount,* or *successive discounts* It is not necessary to find the amount of the discount if we are only interested in the "net price," that is, the amount we have to pay· The net price can be found by subtracting each discount from 100% and then finding the corresponding percentage of the list price.

Examples

(a) Find the net cost of a lot of shirts listed at $168.00 subject to discounts of 25% and 8%.

Solution.

First discount: $100\% - 25\% = 75\%$.
First net price: $\$168 \times \frac{3}{4} = \$126$.
Second discount: $100\% - 8\% = 92\%$.
Second net price: $\$126 \times .92 = \$115.92$.

(b) The list price of a watch in a catalog is $92.00. The catalog quotes discounts of 40%, 10%, 4%. Find the net price.

Solution.

First discount: $100\% - 40\% = 60\%$.
First net price: $\$92 \times .60 = \$55.20$.
Second discount: $100\% - 10\% = 90\%$.
Second net price: $\$55.20 \times .90 = \$49.68$.
Third discount: $100\% - 4\% = 96\%$.
Net price: $\$49.68 \times .96 = \$47.69$. *Ans.*

When two or more discounts are given, the equivalent single rate can be found by multiplying the resulting net rates and subtracting from 100%.

Examples

(a) Find the equivalent single rate for discounts of 30% and 10%.
Solution.

$100\% - 30\% = 70\%$ and $100\% - 10\% = 90\%$.

The equivalent net rate is

$70\% \times 90\% = .70 \times .90 = .63 = 63\%$.

The equivalent single discount rate is

$100\% - 63\% = 37\%$. *Ans.*

(b) Find the equivalent single rate for discounts of 40%, 20%, and 5%.

Solution.

$100\% - 40\% = 60\%$; $100\% - 20\% = 80\%$; $100\% - 5\% = 95\%$.

The equivalent net rate is

$$.60 \times .80 \times .95 = .456 = 45.6\%.$$

The equivalent single discount rate is

$$100\% - 45.6\% = 54.4\%. \ Ans.$$

(c) Find the net price of a gross of pencils listed at $7.20 subject to 15% and 4% off.

Solution.

Equivalent net rate is

$$(100\% - 15\%)(100\% - 4\%) = (.85)(.96) = .816.$$

The net price is

$$\$7.20 \times .816 = \$5.88.$$

Commission

Most salesmen work on commission; that is, their pay is a certain percentage of their sales. The real estate agent's commission is usually 5% on the sale of a house; the automobile salesman may make 10% on the sale of a car; etc. Some salesmen such as clerks in department stores have a fixed salary plus a commission. In most cases a commission is stated in terms of per cent.

Examples

(a) A paint salesman works on a 15% commission and receives an allowance of 8 cents a mile for his car. If he sold $95,000 worth of paints and traveled 35,000 miles during the year, what were his earnings and allowance?

Solution.

Commission: $95,000 \times .15 = \$14,250.00$
Travel allowance: $35,000 \times .08 = \underline{2,800.00}$
Total $\$17,050.00$

(b) A clerk in a department store receives a fixed salary of $62.00 a week and a commission of 5% of the net sales. If a clerk sold $896.00 worth of goods, what was her week's pay?

Solution.

$$\begin{aligned}
\text{Commission:} \quad \$896.00 \times .05 &= \$\ 44.80 \\
\text{Fixed Salary:} \quad &\quad\ \ 62.00 \\
\text{Total} \quad &\ \ \$106.80
\end{aligned}$$

The *net proceeds* is the amount left after the commission has been subtracted from the price received.

Example

Mr. Smith's house was sold by a real estate agent for $18,500 with a commission rate of 5%. How much did Mr. Smith get?

Solution.

Commission: $18,500 × .05 = $925.00.
Net Proceeds: $18,500 − $925 = $17,575.
Mr. Smith got $17,575.00.

Problems XVI.

1. A store advertised "Going Out of Business Sale $33\frac{1}{3}$%, 50%, and 60% off." If you bought items totaling $111.00 at $33\frac{1}{3}$% off, $40.50 at 50% off, and $240.00 at 60% off, how much did you pay?

2. The bill from West Side Lumber Company read: "3% discount for cash." If your purchase totaled $268.75, how much would you save by paying cash?

3. A wholesale catalog states discounts of 40% and 15%. An additional discount of 5% is given if the bill is paid by the 10th of the month. What is the net price of a purchase of $2386.50 if the bill is paid on the 5th of the month?

4. You were offered a job for a fixed salary of $800.00 per month or on a commission of 15% of total sales. During the first six months your sales were $5000.00, $4985.00, $5460.00, $4800.00, $6000.00, $6500.00. Which choice of salary should you have selected?

5. A real estate agent received a commission of 5% of the sale price. During the month of May he sold three houses for $18,750.00, $13,500.00 and $38,900.00. He spent $300.00 on advertising and had additional expenses of $925.00. What were his net earnings?

6. One summer Tom took a job selling on commission for 15% of the total sales. How much did he have to sell in order to earn
   (a) $40 per week?   (b) $50 per week?   (c) $65 per week?

7. A college student sold 630 subscriptions to a $3.75 magazine and earned $787.50. What rate of commission did he earn?

8. A store paid its salesmen a 3% commission and allowed its customers a 5% discount on all bills paid by the 10th of the month. How much did the store net from a gross sale of $12,365.00 if all customers took advantage of the discount?

9. An invoice for $650 reads "less 10%, 8%, and 3%." How much should be paid?

10. A real estate agent receiving a 5% commission sold a house for $44,850.00. What were the net proceeds to the owner?

Chapter IV

GEOMETRY

Introduction

The word *geometry* comes from two Greek words, *gaia* meaning the *earth* and *metrein* meaning *to measure*, and dates back to the days when the subject was primarily used in measuring land. Most historians place the origin of geometry with the Egyptians and Babylonians; however, the practical geometry of today was developed by the Greeks around 300 B.C. Geometric forms and principles are used everywhere. We see them in nature and use them in all our constructions, landscapings, machines, and designs. In our study of geometry we shall concern ourselves with the principles and their applications to everyday life.

Basic Properties and Definitions

Certain terms in geometry are hard to define rigorously. However, we can discuss them and easily have a common understanding. Perhaps the easiest approach is to start with a **geometric solid** or simply a **solid**. Think of a rectangular box. It occupies a certain space, and it is this *space* that is considered to be the solid. Such a solid has *length*, *breadth*, and *thickness*. Thus a solid has three dimensions.

The boundary face of a solid is called a **surface;** for example, the top of the box. A surface has *length* and *breadth* but no thickness. In other words, we are not concerned with the material that the box is made of but only the space and thus consider the surface simply as the area of the top, or the side, or the bottom. Thus a surface has two dimensions.

We now get down to a one dimensional geometric figure which may be thought of as the boundary of a surface, and this is called a **line.** A line has *length* but no breadth or thickness.

Finally, we arrive at a geometric concept which has no dimension and may be described as that which separates one part of a line from an adjoining part; this is called a **point.** A point has neither length, breadth, nor thickness but does have *position*. We identify a point by placing a letter near it.

Strictly speaking, a line has no end, and if we mark it off between two points we are referring to a **line segment**. However, in everyday use the word line nearly always refers to a finite part, and we shall use the word to mean a line segment. A line can be designated by using two letters or one letter placed on top of or beside it. See Figure 2. There are many kinds of lines. A **straight line** is a line having the same direction throughout its whole extent. A **curved line** is a line that is continually changing its direction. A **broken line** is a composition of connected straight lines. Examples are shown in Figure 3.

A ———————— *a* ———————— B

Fig. 2

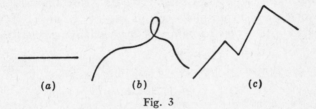

(*a*)   (*b*)   (*c*)

Fig. 3

Two straight lines may intersect or they may not. If they intersect, they meet in one and only one point. If they do not intersect no matter how far they are extended, they are said to be **parallel** and this is indicated by the symbol ∥, which is read "is parallel to." See Figure 4.

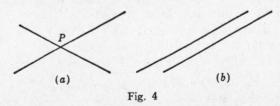

(*a*)   (*b*)

Fig. 4

If any two points in a surface can be connected by a straight line lying entirely in the surface, it is called a **plane surface** or, more often, simply a **plane**. If no part of a surface is a plane, it is called a **curved surface**; an example is the surface of a ball.

The subject of **plane geometry** concerns itself with figures that lie in a plane surface. The subject of **solid geometry** deals with three-dimensional figures.

Angles

Whenever two straight lines meet, they form an angle. However, we can be a little more specific in our definition of an angle. Let us

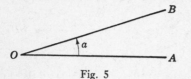

Fig. 5

first consider a **ray** as that part of a straight line on one side of a point on the line. Now when two *rays* are drawn from one point, an **angle** is formed. The rays are the **sides** of the angle and the point at which they meet is called the **vertex** of the angle. Referring to Figure 5, we speak of "the angle *AOB*" or "the angle at *O*." We can also call it the angle *a*. The symbol "∠" is used for the word angle; thus we write ∠*AOB* or ∠*a*.

For a still more rigorous discussion, we start with a line segment *OA* and revolve it about one of its end points, say *O*, until it takes the position *OB*. Then *the revolution of the line has generated the angle*

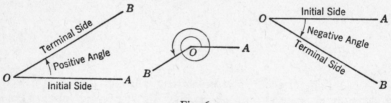

Fig. 6

AOB. The line may be revolved about *O* in a clockwise or counterclockwise direction, and there may be no limit to the number of times it is revolved. This has defined a general angle. If the line is moved in a counterclockwise direction, the angle is *positive;* if moved in a clockwise direction, the angle is *negative.* The line *OA* is called the **initial side** and the line *OB*, the **terminal side.**

Angles are usually measured in the sexagesimal system.

> The **degree** = $\frac{1}{360}$ of one revolution = 1°;
> the **minute** = $\frac{1}{60}$ of one degree = 1′;
> the **second** = $\frac{1}{60}$ of one minute = 1″.

Thus there are 360° in one revolution. If two lines meet such that the four angles which are formed are all equal, then each angle is 360° ÷ 4 = 90° and the angles are called **right angles.** See Figure 7. If an angle is less than a right angle, 90°, it is called an **acute angle.** If it is more than 90° but less than 180°, it is called an **obtuse angle.** If the sum of two angles is 90°, the angles are said to be **complementary**

angles and either one is said to be the **complement** of the other. If the sum of two angles is 180°, they are said to be **supplementary angles**

```
    B                               B
    |                                \
    |                                 \
    |__                                \  ⌒ Obtuse Angle
 90°|  |          A        Acute        _____ A
____|__|____  →   O ___⌒_Angle____ A    O
    |         A                    A
O   |            O
    |
    (a)              (b)                    (c)
```

Fig. 7

and either is said to be the **supplement** of the other. An angle which is 180° is also called a **straight angle** since then *AOB* forms a straight line.

If two lines form a right angle, the lines are said to be **perpendicular** to each other, and this is indicated by the symbol ⊥, which is read "is perpendicular to." A **plumb line** is the line along which a string hangs when fastened at one end and with a weight hanging freely at the other end. This is also called a local **vertical line**. A **horizontal line** is that line which is perpendicular to a vertical line.

A line drawn from a vertex of an angle and dividing the angle into two equal parts is called the **bisector** of the angle.

The instrument for measuring angles is called a **protractor** and consists of a semicircular scale divided into degrees or fractions thereof. See Figure 8. To measure an angle with a protractor, place the line *AB* to coincide with one side of the angle and with the point *O* at the vertex. The other side of the angle (or its extension) will then cross

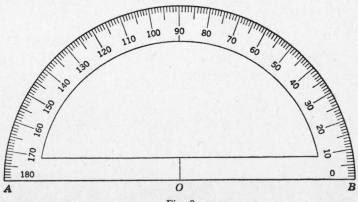

Fig. 8

the scale on the semicircle and the reading off this scale is the measure
of the angle in degrees.

We have already noted that angles are measured in degrees, minutes, and seconds, which are related by the sexagesimal system. In
order to perform arithmetic in this system we have to learn the idea
of carrying and borrowing 60 instead of 10. Thus to change 134 minutes to degrees we divide by 60 to obtain the number of times 60 is
contained in 134. Thus $134' = 2° 14'$ since $134 = 2 \times 60 + 14$.

Examples

(a) Find the angle which is three times as big as $5° 32'$.

Solution.

$$(5° 32')(3) = 15° 96' = 16° 36'.$$

(b) Find the complementary angle of $32° 18'$.

Solution.

The complementary angle of A is $90° - A$. Now to subtract
$32° 18'$ from $90°$ we first borrow $1°$ from $90°$ and write

$$90° = 89° 60'$$

then subtract: $32° 18'$
to obtain: $\overline{57° 42'}$ *Ans.*

(c) Find the supplementary angle of $86° 13' 16''$.

Solution.

The supplementary angle of A is $180° - A$. We write

$$180 = 179° 59' 60''$$

subtract: $86° 13' 16''$
to obtain: $\overline{93° 46' 44''}$ *Ans.*

(d) Find the angle which is one-third of its complementary angle.

Solution: Let x be the angle. Then

$$x = \tfrac{1}{3}(90° - x)$$

or

$$3x + x = 90°$$

so that

$$4x = 90° \quad \text{or} \quad x = 22.5°. \quad Ans.$$

Problems XVII.

1. Find the complementary angles of
   (a) $43°$;   (b) $13° 58'$;   (c) $82° 43' 18''$.

2. Perform the indicated operation:
   (a) $49° 16' 32'' + 62° 43' 45'' + 67° 59' 43''$.
   (b) $16° 32' - 8° 15' 41''$.
   (c) $(22° 11' 18'') \times 5$.
3. What is the size of the two angles formed by bisecting $23° 12' 34''$?
4. If one-half an angle equals two-fifths of the complement of the angle, what is the angle?
5. Find the number of degrees in the angle between the hands of a clock at 1 o'clock, 2 o'clock, 3 o'clock, etc. for each hour.
6. How long does it take the minute hand of a clock to turn through an angle of 90°? An angle of 6°?
7. If a wheel turns at the rate of 120 revolutions per minute, how many degrees does it turn in 2 minutes?
8. If a wagon wheel has twelve equally spaced spokes, what is the angle between two adjacent spokes? How many spokes are there in a right angle?
9. If a spoke of a wheel turns through 480°, how many revolutions does the wheel make?
10. Find the angle which is one-fourth of its complementary angle.

Triangles

The simplest figure which can enclose a plane surface, using straight lines, is a three sided figure which is called a **triangle**. The points at which the lines meet are called **vertices**. Thus a triangle has three sides, three angles, and three vertices.

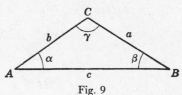

Fig. 9

The individual parts of the triangle are usually labeled according to Figure 9 where the three vertices are A, B, C; the three sides are a, b, c; and the three angles are α (alpha), β (beta), and γ (gamma).

Triangles Classified according to Their Sides:

I. If all three sides are of different length, the triangle is called a **scalene triangle.**

II. If two sides are equal, it is called an **isosceles triangle.**

III. If all three sides are equal, it is called an **equilateral triangle.**

Triangles Classified according to Their Angles:

I. If all three angles are acute angles, the triangle is called an **acute triangle.**

II. If one angle is an obtuse angle, it is called an **obtuse triangle.**

III. If one angle is a right angle, it is called a **right triangle.**

Any triangle which is not a right triangle is also called an **oblique triangle.** There are two special right triangles, namely, the 45°, 45°, 90° triangle and the 30°, 60°, 90° triangle. Such triangles made of various kinds of plastics become drawing instruments for draftsmen, architects, and engineers.

A line from any vertex which is perpendicular to the opposite side and ends there is called an **altitude** of the triangle. Thus each triangle has three altitudes. In some cases the opposite side may have to be

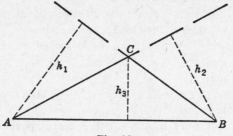

Fig. 10

extended in order for the altitude to meet it. See Figure 10 where the altitudes are labeled h_1, h_2, h_3.

A line drawn from a vertex to the center of the opposite side is called a **median.** See Figure 11.

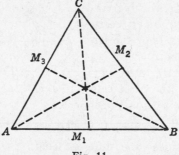

Fig. 11

Two triangles are said to be **equal** or **congruent,** if when placed on top of each other they coincide or fit exactly.

Some Properties of Triangles:

I. The sum of the three angles of any triangle is 180°.

II. The medians of any triangle meet at one point.

III. Two triangles are equal if
    (a) the three sides of one are equal, respectively, to the three sides of the other;
    (b) two sides and the included angle of one are equal, respectively, to two sides and the included angle of the other;
    (c) two angles and the included side of one are equal, respectively, to two angles and the included side of the other.

IV. In an isosceles triangle, the two angles opposite the equal sides are equal.

V. In an equilateral triangle, the three angles are equal and each equals 60°.

VI. The area of a triangle is $\frac{1}{2}$ the base times the altitude.

The Right Triangle

Let us devote a little more time to the right triangle. Consider Figure 12 where we have labeled the important parts of the triangle.

The two sides which form the right angle are also called the *legs* of the right triangle. Since the sum of the angles of a triangle is 180° and since the angle at C is 90°, we have

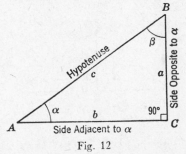

Fig. 12

$$\alpha + \beta = 180° - 90° = 90°.$$

Thus in the right triangle α and β are complementary angles, and if we are given either α or β then we know the size of all the angles in a right triangle. It is also seen that, in Figure 12, the side $\overline{BC} = a$ is the altitude of the triangle.

One of the most important propositions in geometry is the one known as the **Pythagorean proposition.** In specific terms it states:

The square formed on the hypotenuse of a right triangle is equal to the sum of the squares formed on the other two sides.

Again referring to Figure 12 this says that

$$c^2 = a^2 + b^2$$

which we can rewrite in the form

$$c = \sqrt{a^2 + b^2},$$
$$a = \sqrt{c^2 - b^2},$$
$$b = \sqrt{c^2 - a^2}.$$

Now if we are given any two of the sides we can find the third.

Examples

(a) Find the hypotenuse of a right triangle if the two legs are 6 ft. and 4 ft.

Solution. Let $a = 6$ and $b = 4$, then (use Table I)
$$c = \sqrt{36 + 16} = \sqrt{52} = 7.211 \text{ ft.} \quad Ans.$$

(b) Find the side b if $c = 12$ and $a = 9$.

Solution.
$$b = \sqrt{144 - 81} = \sqrt{63} = 7.937. \quad Ans.$$

Since a is an altitude of a right triangle, then by Property VI above we have that the area of a right triangle is given by
$$A = \tfrac{1}{2}ab,$$

where A is the area and a and b are the legs.

The isosceles right triangle has $a = b$ and $\alpha = \beta = 45°$. A per-

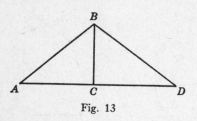

Fig. 13

pendicular dropped from a vertex and meeting the opposite side will divide any triangle into two right triangles. See Figure 13.

If an equilateral triangle is divided into two right triangles we have two 30°, 60°, 90° triangles. If we let the side of the equilateral triangle be s, then we have an interesting set of formulas for the altitude and area. See Figure 14.

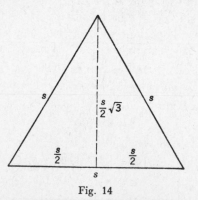

Fig. 14

$$h = \tfrac{1}{2}s\sqrt{3} \quad \text{and} \quad A = \tfrac{1}{4}s^2\sqrt{3}.$$

HANDY ANDY No. 9

Pythagorean Numbers

$2n = a$, $n^2 - 1 = b$, and $n^2 + 1 = c$ yield $c^2 = a^2 + b^2$.
$p^2 - q^2 = a$, $2pq = b$, and $p^2 + q^2 = c$ yield $c^2 = a^2 + b^2$.

n	$2n$	$n^2 - 1$	$n^2 + 1$	q	p	$p^2 - q^2$	$2pq$	$p^2 + q^2$
2	4	3	5	2	3	5	12	13
3	6	8	10	3	4	7	24	25
4	8	15	17	4	5	9	40	41
5	10	24	26	5	6	11	60	61
6	12	35	37	6	7	13	84	85
etc.						etc.		

Problems XVIII.

1. One of the equal angles of an isosceles triangle is 32° 18′; what are the sizes of the other two angles?
2. In a right triangle $c = 37$ and $a = 12$; what is b?
3. Find the area of the following triangles:
   (a) base = 12, altitude = 3;
   (b) $a = b = c = 7$;
   (c) right triangle with $c = 25$ and $a = 7$.
4. Two boats start from the same point and at the same time. One sails due east at the rate of 8 miles per hour, and the other sails due north at the rate of 15 miles per hour. How far apart are they at the end of two hours? Remember that the distance equals the speed multiplied by the time and that the given directions form a right angle.
5. A "rule of thumb" for finding the diagonal of a square is "multiply the length of a side by 10, subtract 1% of this product, and then divide by 7." Test the accuracy of this rule if the side of the square is 5 inches. (The diagonal is the hypotenuse of an isosceles right triangle.)
6. Draw a triangle whose sides are the Pythagorean numbers 10, 24, and 26 (let 1 be $\frac{1}{4}$ inch.) Join the midpoints of these sides forming 4 triangles. What kinds of triangles are formed and what is the area of one of them?
7. A man swims at right angles to a river current. If he swims at the rate of 3 mph (miles per hour), and the current is 7 mph, find

the rate at which he is moving.   (He will be moving along the hypotenuse of a right triangle.)

8. By using the area formula for an equilateral triangle show that the equilateral triangle made on the hypotenuse of a right triangle, whose legs are 3 and 4, is equal to the sum of the equilateral triangles made on the other two sides.

9. Flying in an airplane at constant altitude of 5,000 ft. and in a straight line in the direction A to B to C, the pilot observed a beacon at point Q on the ground and measured the angle BAQ to be 40°.   When he reached the point B he found the distance AB to be 9 miles and measured the angle CBQ to be 80°.   What is the distance from B to Q?   Remember that the angle $ABC = 180°$.

10. The perimeter (sum of the sides) of a triangular flower bed is 25 feet.   If the second side is $\frac{2}{3}$(the first side minus 5) and the third side is $\frac{1}{2}$(the first side plus 2), what are the lengths of the sides?

11. Find the dimensions of a right triangle if the hypotenuse is 17 inches and one leg is 7 inches longer than the other.

12. If the area of a triangle is 11 square inches and the base is 20 inches more than twice the altitude, find the altitude and base.

Polygons

The triangle is a polygon with the smallest number of sides, so we have already met the polygon.   More precisely, a **polygon** is a plane surface bounded by any number of straight lines.   The parts of the polygon are:

a **side,** which is any one of the straight lines;
a **vertex,** which is the point where any two of the lines meet;
the **perimeter,** which is the sum of the lengths of the sides;
an **angle,** which is formed by any two sides;
a **diagonal,** which is a line joining any two nonadjacent vertices.

Polygons are usually classified according to the number of sides.   The most common ones are:

Number of Sides	Name
3	triangle
4	quadrilateral
5	pentagon
6	hexagon
8	octagon

A **regular polygon** is one whose sides are all equal and whose angles are all equal. All polygons can be divided into a number of triangles by drawing all possible diagonals from one vertex. See Figure 15.

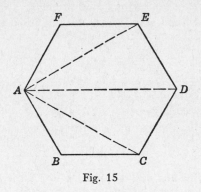

Fig. 15

$$ABCDEF = \triangle AEF + \triangle AED + \triangle ACD + \triangle ABC.$$

There are four common quadrilaterals.

A **parallelogram** is a quadrilateral whose opposite sides are parallel.
A **rectangle** is a parallelogram whose angles are right angles.
A **square** is a rectangle whose sides are all equal.
A **trapezoid** is a quadrilateral with only two parallel sides.

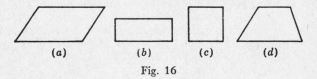

(a)   (b)   (c)   (d)

Fig. 16

There are some interesting properties connected with the diagonals of these special quadrilaterals.

   I. A diagonal of a square divides the square into two equal isosceles right triangles.
  II. A diagonal of a parallelogram divides it into two equal triangles.
 III. The diagonals of a parallelogram bisect each other.
 IV. The diagonals of a rectangle are equal.
  V. The diagonals of a square are perpendicular.
 VI. The diagonal of a square of side s equals $s\sqrt{2}$.

One of the important practical problems connected with plane figures is that of finding the area enclosed. We already have a formula for the area of a triangle, and we could find the area of polygons by first dividing them into triangles and then finding the sum of the areas of the triangles. However, it is more convenient to have formulas which obtain the area directly. For the four special quadrilaterals these are given in the following table where A is the area, a is the altitude, b is a side, and (for the trapezoid) B and b are the parallel sides.

Quadrilateral	Area
Square	$A = a^2$
Rectangle	$A = ab$
Parallelogram	$A = ab$
Trapezoid	$A = \frac{1}{2}a(B + b)$

For the trapezoid the altitude is the perpendicular length between the two parallel sides.

Examples

(a) Find the area of a trapezoid whose altitude is 4 in. and the parallel sides are 12 in. and 7 in.

Solution.

Given: $a = 4$, $B = 12$, and $b = 7$.
Then $A = \frac{1}{2}(4)(12 + 7) = 38$ sq. in. *Ans.*

(b) Find the number of square feet in the floor of the room shown in Figure 17(a).

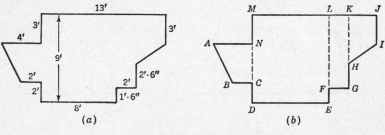

Fig. 17

Solution.

Label the vertices as shown in Figure 17(b) and draw the lines CN, $FL \perp MJ$, and $HK \perp MJ$. We have then divided the room into two trapezoids and two rectangles.

Trapezoid $ABCN$:

$B = AN = 4'$, $b = BC = 2'$, $a = CN = 9' - (2' + 3') = 4'$.
$$A_1 = \tfrac{1}{2}(4)(4 + 2) = 12 \text{ sq. ft.}$$

Rectangle $DELM$:

$a = 9'$ and $b = 8'$.
$$A_2 = (9)(8) = 72 \text{ sq. ft.}$$

Rectangle $FGKL$:

$a = 2'$ and $b = 9' - 1'\,6'' = 7\tfrac{1}{2}'$.
$$A_3 = (2)(7\tfrac{1}{2}) = 15 \text{ sq. ft.}$$

Trapezoid $HIJK$:

$B = 9' - (1'\,6'' + 2'\,6'') = 5'$, $b = 3$, $a = 13' - (8' + 2') = 3'$.
$$A_4 = \tfrac{1}{2}(3)(5 + 3) = 12 \text{ sq. ft.}$$

Then the total area is

$$A = A_1 + A_2 + A_3 + A_4 = 12 + 72 + 15 + 12 = 111 \text{ sq. ft.}$$

Ans.

Problems XIX.

1. What is the price of a 9 by 12 ft. rectangular rug if the price is $1.25 per square foot?
2. Find the cost of sodding a rectangular lawn 35 ft. wide and 48 ft. long at 60 cents a square yard? (There are 9 square feet in 1 square yard.)
3. Find the total area contained in Figure 18.

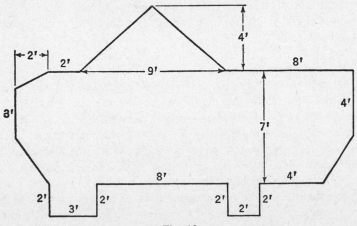

Fig. 18

4. Many nuts for bolts are made in the shape of regular hexagons. The diagonal which divides the hexagonal nut into two equal parts is called the distance "across corners." The distance between two parallel sides is called the distance "across the flats" and determines the size of the wrench to be used. Draw a regular hexagon and a diagonal across the corners. From the midpoint of this diagonal draw lines to all vertices and satisfy yourself that the hexagon is composed of six equilateral triangles and that the distance across corners is $2s$, where s is a side. Now check the fact that the distance across the flats is $2h$ where h is the altitude of one of the triangles. Using the formula $h = \frac{1}{2}s\sqrt{3}$, show that the distance across corners is approximately 1.15 times the distance across the flats.

5. A painting measuring 10 inches by 18 inches was put in a frame of uniform width and it was found that the area of the frame was 60 square inches more than the area of the painting. Find the dimensions of the frame.

6. What is the area of a square whose perimeter is 32 inches? What is the length of the diagonal?

7. It is decided to put a walk which is 3 feet wide through the middle of a rectangular garden which is 20 feet by 30 feet. How much area is left for gardening if the walk runs the length of the garden? How much if the walk runs the width of the garden?

8. What is the length of a side of a square if the length of the diagonal is $32\sqrt{2}$ feet?

9. The open end of a dog house is in the shape of an isosceles triangle placed on a rectangle. If the rectangle is $1\frac{1}{2}$ feet high and 2 feet wide and the altitude of the triangle is 1 foot, what is the area of the open end?

Similar Figures

There are three important ways in which geometric figures can be compared. We have already discussed one method when we said that two triangles were **congruent** if they could be made to fit each other exactly, or in other words if they had the same shape and the same size. If two figures have the same shape but are different in size, then they are said to be **similar** figures. The third comparison is the case when two figures have the same size but are different in shape; they are then said to be **equivalent** figures. The concept of similarity is very useful, for if two geometric figures are similar then their corresponding angles are equal and their corresponding sides are proportional. These two properties are often taken as the definition for similarity and lead to the following property:

Any two **regular polygons** *with the same number of sides are similar to each other.*

Figure 19 shows examples of similar polygons.

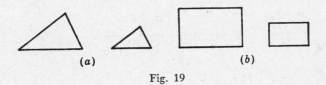

(a)   (b)

Fig. 19

Since similar figures have the corresponding sides proportional, it is possible to calculate unknown parts by means of ratios and proportions (see Section 21).

Examples

(a) Given two similar triangles ABC and $A'B'C'$. If the sides of $\triangle ABC$ are $a = 5$, $b = 7$, and $c = 9$ and if $a' = 15$, find b' and c'.

Solution.

Since the triangles are similar the corresponding sides are proportional. Thus

$$a:a'::b:b' \quad \text{or} \quad \frac{5}{15} = \frac{7}{b'}$$

and solving for b' we have $b' = 21$.

Similarly, $\dfrac{5}{15} = \dfrac{9}{c'} \quad$ or $\quad c' = 27$.

(b) If at the same time that the shadow of a telephone pole measures 54 ft., a man 6 ft. tall standing nearby casts a shadow 9 ft. long, how high is the pole?

Solution.

Form the two similar triangles of Figure 20, p. 90.

Then $\dfrac{9}{54} = \dfrac{6}{x} \quad$ or $\quad x = 36$ ft. *Ans.*

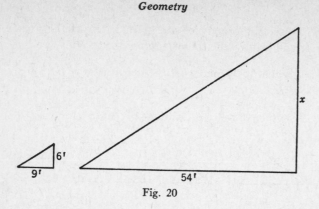

Fig. 20

Trigonometry of Triangles

If two right triangles have an acute angle of one equal to an acute angle of the other, the triangles are similar and the sides are proportional. Let us consider an angle and form the similar right triangles shown in Figure 21. Since they all have $\angle \alpha$ in common, they are similar and we have the proportions

$$\frac{B'C'}{AC'} = \frac{BC}{AC} = \frac{B''C''}{AC''}$$

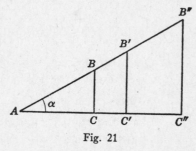

Fig. 21

or, in words, the ratio of the opposite side to the adjacent side in each triangle is the same. This leads to the very useful concept of the trigonometry of triangles, which is a branch of mathematics that is concerned with the solutions of triangles. We shall limit our discussion to the very elementary part of trigonometry which is based on the right triangle. We start by naming the six trigonometric functions.

sine of α;	abbreviated: sin α;
cosine of α;	abbreviated: cos α;
tangent of α;	abbreviated: tan α;
cotangent of α;	abbreviated: cot α;
secant of α;	abbreviated: sec α;
cosecant of α;	abbreviated: csc α;

If α is an acute angle of a right triangle, these six trigonometric functions are defined to be

$$\sin\ \alpha = \frac{\text{opposite side}}{\text{hypotenuse}}\ ; \quad \cot\ \alpha = \frac{\text{adjacent side}}{\text{opposite side}}\ ;$$

$$\cos\ \alpha = \frac{\text{adjacent side}}{\text{hypotenuse}}\ ; \quad \sec\ \alpha = \frac{\text{hypotenuse}}{\text{adjacent side}}\ ;$$

$$\tan\ \alpha = \frac{\text{opposite side}}{\text{adjacent side}}\ ; \quad \csc\ \alpha = \frac{\text{hypotenuse}}{\text{opposite side}}$$

Consider now the right triangle of Figure 22. By the above definitions we see that

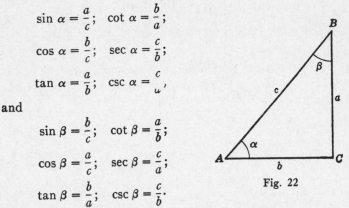

$$\sin\ \alpha = \frac{a}{c}\ ; \quad \cot\ \alpha = \frac{b}{a}\ ;$$

$$\cos\ \alpha = \frac{b}{c}\ ; \quad \sec\ \alpha = \frac{c}{b}\ ;$$

$$\tan\ \alpha = \frac{a}{b}\ ; \quad \csc\ \alpha = \frac{c}{a}\ ,$$

and

$$\sin\ \beta = \frac{b}{c}\ ; \quad \cot\ \beta = \frac{a}{b}\ ;$$

$$\cos\ \beta = \frac{a}{c}\ ; \quad \sec\ \beta = \frac{c}{a}\ ;$$

$$\tan\ \beta = \frac{b}{a}\ ; \quad \csc\ \beta = \frac{c}{b}\ .$$

Fig. 22

Since these ratios are the same for a given angle no matter what the size of the respective sides, there is one unique number for each trigonometric function of a given angle. These can thus be published in a table and we have extensive tables; * however, for our purpose we shall consider only the simple table printed as Table III in the back of this book. If greater accuracy is desired the reader is referred to the book mentioned in the footnote on this page.

A triangle has six parts, namely, three angles and three sides. In a right triangle one angle is 90°; thus it is only necessary to have given two sides or an acute angle and a side in order to compute the remaining parts of a right triangle. The finding of the unknown parts is called the *solution* of the triangle.

Examples

(a) Solve the right triangle if $\alpha = 23°$ and $a = 5$.

Solution.

By Formula: $\quad \sin\ \alpha = \dfrac{a}{c}\quad$ or $\quad c = \dfrac{a}{\sin\ \alpha}\ .$

* See K. L. Nielsen and J. H. Vanlonkhuyzen, *Plane & Spherical Trigonometry* (New York, Barnes & Noble, Inc., 1954).

From Table III: sin 23° = .3907.

$$c = \frac{5}{.3907} = 12.8. \quad Ans.$$

By Formula: $\tan \alpha = \dfrac{a}{b}$ or $b = \dfrac{a}{\tan \alpha}$.

From Table III: tan 23° = .4245.

$$b = \frac{5}{.4245} = 11.8. \quad Ans.$$

Since $\alpha + \beta = 90°$, $\beta = 90° - 23° = 67°$. *Ans.*

(b) If the angle from the horizontal ꞈt the foot of a building to the top of the building measures to be 50° at a distance 200 ft. from the building, how high is the building?

Solution.

Draw Figure 23.   Then

$$\tan 50° = \frac{x}{200}$$

or

$$x = 200(1.1918)$$
$$= 238.36 \text{ ft.} \quad Ans.$$

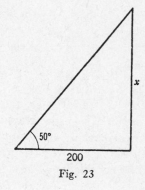

(c) Solve the right triangle when $a = 3$ and $b = 4$.

Fig. 23

Solution.

From the formulas we have

$$\tan \alpha = \frac{a}{b} = \frac{3}{4} = .75.$$

Now to find the angle whose tangent is .75 we search Table III in the "tan" column until we find the number closest to .75, which is .7536, and in the left column we find $\alpha = 37°$. This is the answer to the nearest degree. If greater accuracy is desired, we should employ more extensive tables.

Since $\alpha + \beta = 90°$, $\beta = 90° - 37° = 53°$. From the Pythagorean proposition we have

$$c = \sqrt{a^2 + b^2} = \sqrt{9 + 16} = 5. \quad Ans.$$

HANDY ANDY No. 10						
Trigonometric Functions of Special Angles						
Angle	sin	cos	tan	cot	sec	csc
0°	0	1	0	—	1	—
30°	$\frac{1}{2}$	$\frac{1}{2}\sqrt{3}$	$\frac{1}{3}\sqrt{3}$	$\sqrt{3}$	$\frac{2}{3}\sqrt{3}$	2
45°	$\frac{1}{2}\sqrt{2}$	$\frac{1}{2}\sqrt{2}$	1	1	$\sqrt{2}$	$\sqrt{2}$
60°	$\frac{1}{2}\sqrt{3}$	$\frac{1}{2}$	$\sqrt{3}$	$\frac{1}{3}\sqrt{3}$	2	$\frac{2}{3}\sqrt{3}$
90°	1	0	—	0	—	1

Problems XX.

1. Find the values of the following functions:
   (a) sin 12°.   (b) cos 64°.   (c) tan 59°.

2. Find the following angles:
   (a) sin α = .6157.   (b) cos α = .8090.   (c) tan α = 1.2349.

3. Solve the following right triangles (obtaining the angles to the nearest degree):
   (a) Given $a = 7$ and $\alpha = 28°$.
   (b) Given $a = 12$ and $b = 5$.
   (c) Given $\beta = 45°$ and $c = \sqrt{2}$.

4. To find the distance x across a river a surveyor establishes a triangle on his side of the river with two sides equal to 5 feet and 10 feet. He then establishes a similar triangle with corresponding sides of $(x + 5)$ feet and 60 feet. Find x.

5. If a force is acting at an angle to the horizontal, it can be decomposed into its *horizontal* and *vertical* components by considering it to be acting along the hypotenuse of a right triangle. Let the angle be α and the force be F, then the horizontal component is $b = F \cos \alpha$ and the vertical component is $a = F \sin \alpha$. Find the components of a force of 300 pounds acting at 43° with the horizontal.

6. Under the conditions of problem 5, what is the force if the horizontal component is 15 pounds and the angle is 60°?

7. From the top of a lighthouse 200 feet above a lake, the keeper spots a boat sailing directly towards him. He observes the angle between the horizontal and the boat to be 6° and then later to be 14°. Form two right triangles with the height of the lighthouse and the distance from the foot of the lighthouse to the boat as the legs. The two measured angles will be equal to the angles at the boat. Find the distance the boat has traveled between the two observations.

Trigonometric Identities

The trigonometric functions discussed in the preceding section have a number of interesting properties of which there is a group known as the fundamental identities. There are eight of these which we shall express in terms of the angle θ (theta).

The Reciprocal Relations.

1. $\csc \theta = \dfrac{1}{\sin \theta}$; 　 2. $\sec \theta = \dfrac{1}{\cos \theta}$; 　 3. $\cot \theta = \dfrac{1}{\tan \theta}$.

The Quotient Relations.

4. $\tan \theta = \dfrac{\sin \theta}{\cos \theta}$; 　 5. $\cot \theta = \dfrac{\cos \theta}{\sin \theta}$.

The Pythagorean Relations.

6. $\sin^2 \theta + \cos^2 \theta = 1$;
7. $\tan^2 \theta + 1 = \sec^2 \theta$;
8. $1 + \cot^2 \theta = \csc^2 \theta$.

These can be proved directly from the definitions; for example, since by the definitions

$$\csc \alpha = \frac{c}{a} \quad \text{and} \quad \sin \alpha = \frac{a}{c}$$

we see that

$$\csc \alpha = \frac{c}{a} = \frac{1}{\dfrac{a}{c}} = \frac{1}{\sin \alpha}$$

by dividing both numerator and denominator by c at step 2.

We see from the reciprocal relations that we could really limit our study of these functions to that of the sine, cosine, and tangent, for the other three can be expressed in terms of them.

The fundamental identities are used to prove other trigonometric identities, of which there are a large number.

Example

Prove the identity $\tan \alpha + \cot \alpha = \sec \alpha \csc \alpha$.

Solution. We shall alter only the left member.
Using the *Quotient relations* (4) and (5):

$$\frac{\sin \alpha}{\cos \alpha} + \frac{\cos \alpha}{\sin \alpha} =$$

Simplifying: $\dfrac{\sin^2 \alpha + \cos^2 \alpha}{\cos \alpha \sin \alpha} =$

Using the *Pythagorean relation* (6):

$$\frac{1}{\cos \alpha \sin \alpha} = \qquad \text{or} \qquad \frac{1}{\cos \alpha} \frac{1}{\sin \alpha} =$$

Using the *Reciprocal relations* (1) and (2):

$$\frac{1}{\cos \alpha} \frac{1}{\sin \alpha} = \sec \alpha \csc \alpha.$$

Here are two suggestions for proving trigonometric identities:

1. All functions may be expressed in terms of sines and cosines.
2. If one member involves only one function, express everything on the other side in terms of this function.

Problems XXI.

1. Prove the following identities.

   (a) $\dfrac{\sin \alpha}{\csc \alpha} + \dfrac{\cos \alpha}{\sec \alpha} = 1.$

   (b) $\dfrac{\sec \alpha}{\cot \alpha + \tan \alpha} = \sin \alpha.$

   (c) $\cot \alpha \sin \alpha \sec \alpha = 1.$

   (d) $\cos \alpha + \tan \alpha \sin \alpha = \sec \alpha.$

   (e) $\tan \alpha \sin \alpha + \cos \alpha - \sec \alpha = 0.$

The Circle

The circle occurs often in practical mathematics. A **circle** is a plane figure bounded by a curved line every point of which is the same dis-

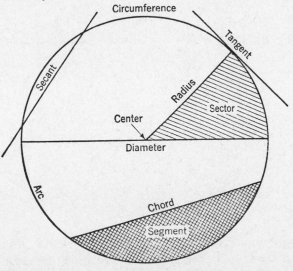

Fig. 24

tance from another point, called the **center**. Some important parts of a circle are shown in Figure 24, p. 95. The definitions are:

The **circumference** is the length of the curved line.

Any part of the circumference is called an **arc**.

A line drawn through the center and terminating in the circumference is called a **diameter**.

The diameter divides the circle into two equal parts called **semicircles**.

Half the diameter is called a **radius**; it is the distance from the center to the curved line.

The straight line joining the ends of an arc is called a **chord,** and the chord is said to **subtend** its arc.

A straight line which touches the circle at one point only is called a **tangent**.

If a straight line cuts a circle at two points, it is called a **secant**.

The area bounded by an arc and a chord is called a **segment**.

The area bounded by two radii and an arc is called a **sector** (the pie-shaped piece).

Figure 25 illustrates the two types of angles. A **central angle** is an angle with its vertex at the center of the circle ($\angle DOE$); and an **inscribed angle** is an angle with its vertex on the circumference of the circle ($\angle ABC$). These angles are said to **intercept** the arcs between their sides.

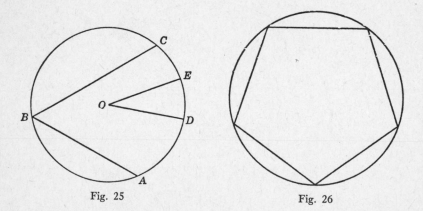

Fig. 25   Fig. 26

If a polygon is inside a circle and has its vertices on the circumference, it is said to be **inscribed in** the circle, and the circle is said to be **circumscribed about** the polygon (Figure 26). A polygon is **cir-**

cumscribed about a circle and the circle is inscribed in a polygon if the sides of the polygon are all tangent to the circle.

There are many interesting properties connected with circles, and for a thorough study the reader should consult a book on plane geometry.* We shall list a few of the most practical ones.

I. In the same circle or equal circles, chords that are the same distance from the center are equal.

II. A radius drawn to the center of a chord is perpendicular to the chord and bisects the arc subtended by the chord.

III. A radius drawn to the point of contact of a tangent is perpendicular to the tangent.

IV. If two circles are tangent to each other, the straight line joining their centers passes through the point of tangency.

V. An inscribed angle is measured by half its arc. Thus an angle inscribed in a semicircle is a right angle.

VI. A central angle has the same measure as its arc.

Many practical problems are concerned with the measurements of a circle, and basic to the measurements is the fact that the *ratio of the circumference to its diameter is a constant*, no matter what the size of the circle. Unfortunately this constant is not an integer. In mathematics the ratio is represented by the Greek letter π (pi) and much effort has been spent to find the value of this ratio. It can be found to a large number of decimal points and has been evaluated to more than 200 significant digits by electronic calculators. For most practical problems the approximation 3.1416 is sufficient, but for those who are interested here is the approximate value to 20 decimal places:

$$3.14159265358979323846.$$

We are now in a position to write some measurement formulas. Let C denote the *circumference;* r, the *radius;* d, the *diameter;* and A, the *area*. Then

$$C = \pi d = 2\pi r,$$
$$A = \tfrac{1}{2}Cr = \pi r^2 = \tfrac{1}{4}\pi d^2.$$

These formulas are used to calculate the parts of the circle.

Examples

(a) Find the circumference and the area of a circle whose radius is 5 inches.

* See also M. Horblit and K. L. Nielsen, *Problems in Plane Geometry* (New York, Barnes & Noble, Inc., 1947).

Solution.

By the formula: $C = 2\pi r = 2(3.1416)(5) = 31.416$ in.

By the formula: $A = \pi r^2 = 3.1416(5)^2 = 78.54$ sq. in.

(b) Find the radius and circumference of a circle whose area is 113.0976 sq. in.

Solution.

By the formula we have $A = \pi r^2$ or

$$r = \sqrt{A \div \pi}.$$

Therefore

$$r = \sqrt{\frac{113.0976}{3.1416}} = \sqrt{36} = 6 \text{ in.}$$

Then

$$C = 2\pi r = 2(3.1416)(6) = 37.6992 \text{ in.}$$

HANDY ANDY No. 11

Multiples of π

No.	Value	Log *	No.	Value	Log *
π	3.1416	0.4971	$\frac{1}{\pi}$	0.3183	$9.5029 - 10$
2π	6.2832	0.7982	$\frac{1}{2}\pi$	1.5708	0.1961
3π	9.4248	0.9743	$\frac{1}{3}\pi$	1.0472	0.0200
4π	12.5664	1.0992	$\frac{1}{4}\pi$	0.7854	$9.8951 - 10$
5π	15.7080	1.1961	$\frac{1}{5}\pi$	0.6283	$9.7982 - 10$
6π	18.8496	1.2753	$\frac{1}{6}\pi$	0.5236	$9.7190 - 10$
7π	21.9912	1.3422	$\frac{1}{7}\pi$	0.4488	$9.6521 - 10$
8π	25.1328	1.4002	$\frac{1}{8}\pi$	0.3927	$9.5941 - 10$
9π	28.2744	1.4514	$\frac{1}{9}\pi$	0.3491	$9.5429 - 10$
π^2	9.8696	0.9943	$\sqrt{\pi}$	1.7725	0.2486

The *area of a sector of a circle* is given by the formula

$$A = \frac{\theta}{360} \pi r^2$$

where θ is the number of degrees in the angle of the sector.

* The columns headed "Log" give the logarithms of π and its multiples. For the meaning and use of logarithms, see Chapter VII.

Example

Find the area of the sector which has a central angle of 45° and a radius of 10 inches.

Solution.

Here we have $\theta = 45°$ and $r = 10$ so that a substitution into the formula yields

$$A = \tfrac{45}{360}(3.1416)(10^2) = \tfrac{1}{8}(314.16)$$
$$= 39.27 \text{ sq. in. } Ans.$$

Frequently, many problems can be solved by what are called handbook techniques. This amounts to having a handbook that lists numbers which calculate a certain portion of a formula on the basis of unit measurement. These numbers then become multiplying factors. Such a technique can be applied nicely to regular polygons since their sides are equal. The following table lists multiplying factors for regular polygons.

REGULAR POLYGONS

No. of Sides	Polygon	Area	Radius of Circumscribed Circle	Radius of Inscribed Circle	Side if Radius of Circumscribed Circle = 1
		if Length of Side = 1			
3	triangle	0.433013	0.5773	0.2887	1.7321
4	square	1	0.7071	0.5000	1.4142
5	pentagon	1.720477	0.8056	0.6882	1.1756
6	hexagon	2.598076	1	0.8660	1
7	heptagon	3.633912	1.1524	1.0383	0.8677
8	octagon	4.828427	1.3066	1.2071	0.7653
9	nonagon	6.181824	1.4619	1.3737	0.6840
10	decagon	7.694209	1.6180	1.5388	0.6180
12	dodecagon	11.196152	1.9319	1.8660	0.5176

I. *To find the area, given the length of a side, multiply the square of the side by the corresponding number given in the column "Area."*

Example

Find the area of a hexagon having sides of 9 feet.

Solution.

In the area column opposite hexagon we find the number 2.598076. Thus

$$A = (9^2)(2.598076) = 210.444156 \text{ sq. ft. } Ans.$$

II. *To find the radius of the circumscribed circle, given the length of the side, multiply the length of the side by the corresponding number given in the fourth column.*

Example

Find the radius of the circumscribed circle of a pentagon whose sides are 3 inches.

Solution.

The number in the fourth column for a pentagon is 0.8056. Thus

$$R = (0.8056)(3) = 2.4168 \text{ inches. } Ans.$$

III. *To find the radius of the inscribed circle, given the length of the side, multiply the length of the side by the corresponding number given in the fifth column.*

Example

Find the radius of the inscribed circle of an octagon whose sides are 4 inches.

Solution.

The number in the fifth column for an octagon is 1.2071. Thus

$$r = (1.2071)(4) = 4.8284 \text{ in. } Ans.$$

IV. *To find the length of a side of a polygon that can be inscribed in a circle of given radius, multiply the radius by the corresponding number given in the sixth column.*

Example

Find the length of the sides of a square inscribed in a circle whose radius is 5 inches.

Solution.

The number in the sixth column corresponding to a square is 1.4142. Thus

$$s = (1.4142)(5) = 7.071 \text{ in. } Ans.$$

Problems XXII.

1. Find the circumference and the area of the following circles.
   (a) $r = 8$.   (b) $d = 12$.   (c) $r = 6.2$.

2. Find the radius, diameter, and area of the circles whose circumferences are   (a) 22.   (b) 16.   (c) 11.4.

3. What is the area of a sector of a circle whose radius is 4 inches if the central angle is 30°?

4. Find the area of a ring formed by two concentric (having the same center) circles of radius 6 in. and 4 in., respectively.

5. Find the area of a triangle inscribed in a semicircle of radius 5 if one of the shorter sides of the triangle is equal to 6.

6. Find the area of each of the following polygons:
   (a) pentagon of side 3 inches.
   (b) nonagon of side 7 inches.
   (c) hexagon of side 2.5 inches.

7. Find the radii of the circumscribed and inscribed circles of the polygons of Problem 6.

8. Find the length of the sides of all polygons listed in the table of this section if the radius of the circumscribed circle is 2.5 inches.

9. A circular flower bed 16 feet in diameter is surrounded by a walk 3 feet wide. What is the area of the walk?

10. Let the equator be a circle and the radius of the earth be 3960 miles. How much farther would a plane flying at a constant altitude of 5280 feet travel than a man staying on the surface if both go around the world at the equator?

11. The perimeter of a semicircular flower bed is 15.4248 feet. What is its area?

12. The side of a regular hexagon which is inscribed in a circle is equal to the radius of the circle. What is the difference in area between a circular swimming pool of radius 15 feet and that of a hexagonal swimming pool if the hexagon is inscribed in the same circle?

13. What is the circumference of a bicycle wheel if the diameter is 26 inches?

14. The minute hand of a grandfather's clock is 8 inches from the pivot point to the tip. How far does the tip travel when the hand moves for 36 minutes?

15. A race track is made in the form of a rectangle with two semicircles on the shorter sides. What is the area of the "infield" if the rectangle measures $\frac{3}{8}$ mile by $\frac{1}{8}$ mile? What is the length of the track?

Solid Geometric Forms

We shall now turn our attention to three-dimensional geometric configuration, the geometric solids. Since we live in a three-dimensional space, these solids are already familiar to us through everyday experiences. We shall first illustrate the more common solids and then turn to their measurements in the next section.

A **prism** is a solid whose **bases** are parallel polygons and whose **sides** are parallelograms. In Figure 27 we have a prism; the **bases** are

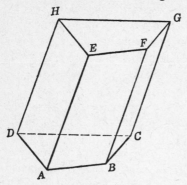

Fig. 27

ABCD and *EFGH*. The **sides** (also called *faces*) are *ABFE*, *BCGF*, etc. The lines *AB*, *BC*, *CD*, *AD* and *EF*, *FG*, *GH*, *EH* are called **base edges,** and *AE*, *BF*, etc. are called **lateral edges.** If the lateral edges are perpendicular to the bases of the prism, the prism is a **right prism.** If bases of a right prism are rectangles, it is called a **rectangular solid** and, furthermore, if all six faces are squares, it is called a **cube.** See Figure 28.

Cube

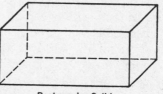

Rectangular Solid

Fig. 28

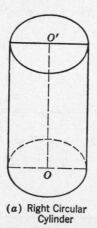

(*a*) Right Circular Cylinder

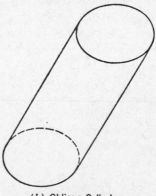

(*b*) Oblique Cylinder

Fig. 29

A **right circular cylinder** is a solid formed by revolving a rectangle about one of its sides as an axis. See Figure 29(a). For this cylinder, the two bases are circles and the line joining the centers of the bases is called the **axis** of the cylinder. Since the axis is perpendicular to the bases, it is equal to the **altitude** of the cylinder. Cylinders may also have other closed curves as bases and may be oblique as shown in Figure 29(b).

A **pyramid** is a solid whose base is a polygon and whose sides are triangles with their vertices at a common point which is called the **vertex** of the pyramid. A **right pyramid** is a pyramid whose base is a regular polygon and the sides are equal isosceles triangles. See Figure 30. The **altitude** of the pyramid is the line from the vertex

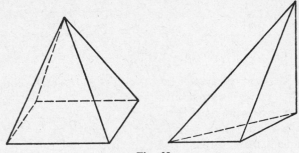

Fig. 30

perpendicular to the base. The **slant height** of a right pyramid is the line drawn from the vertex to the center of one edge of the base. A **lateral edge** is the line in which two sides meet.

A **circular cone** is a solid whose base is a circle and whose lateral surface tapers to a point, called the **vertex.** See Figure 31. If the

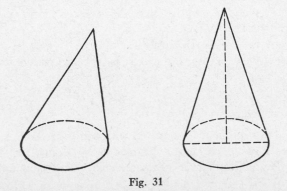

Fig. 31

base is a circle and the axis is perpendicular to the base, the cone is called a **right circular cone**. The **altitude** of a cone is the line from the vertex perpendicular to the base. The **slant height** is a straight line from the vertex to the circumference of the base.

If the top of a pyramid or a cone is cut off by a plane parallel to **the** base, the remaining part is called a **frustum** of a pyramid or a cone. See Figure 32. The **altitude** of a frustum is the length of the per-

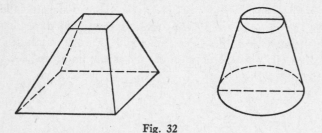

Fig. 32

pendicular between the bases, and the **slant height** is the shortest line between the perimeters of the two bases.

A **sphere** is a solid bounded by a curved surface, every point of which is equally distant from a point within, called the **center**. A straight line passing through the center and ending in the surface is called a **diameter,** and a line from the center to the surface is called a **radius**. See Figure 33. If a sphere is cut by a plane, the section formed is a circle. If the plane passes through the center, we have a **great circle;** otherwise we have a **small circle**.

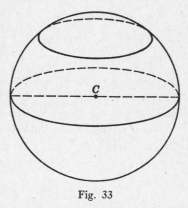

Fig. 33

The circumference of a great circle is the circumference of the sphere.

Surfaces and Volumes of Solids

Most practical problems connected with solids are concerned with measurements of the surfaces and volumes. We shall list formulas which are derived in a formal geometry course and show how to use them.

Let us use the following letters:

> S for the lateral area,
> T for the total area,
> A for the area of each base,
> p for the perimeter of the base,
> h for the altitude,
> a for an edge, and
> V for the volume.

I. *Prism.* For a right prism we have (where the lateral area is the area of its sides not including the two bases):

$S = ph$	$V = Ah$
$T = ph + 2A$	$V = a^3$ (for a cube)
$T = 6a^2$ (for a cube)	

II. *Cylinder.* For a right circular cylinder with d being the diameter and r, the radius of the base, we have

$$S = \pi dh = 2\pi rh, \quad V = Ah = \pi r^2 h,$$
$$T = 2\pi rh + 2A = 2\pi r(h + r).$$

III. *Pyramid and Cone.* For a pyramid and cone with s being the slant height we have

$$S = \tfrac{1}{2}ps,$$
$$T = \tfrac{1}{2}ps + A,$$
$$V = \tfrac{1}{3}Ah.$$

IV. *Frustum.* For a frustum we have

$$S = \tfrac{1}{2}(P + p)s,$$
$$T = \tfrac{1}{2}(P + p)s + B + b,$$
$$V = \tfrac{1}{3}h(B + b + \sqrt{Bb}),$$

where P and B refer to the perimeter and area of the lower base and p and b refer to the perimeter and area of the upper base.

V. *Sphere.* For the sphere we have

$$S = T = 4\pi r^2 \quad \text{and} \quad V = \tfrac{4}{3}\pi r^3.$$

Examples

(a) Find the total surface of a rectangular box measuring 2′ by 6′ by 3′.

Solution.

Let the base be $2' \times 6'$; then the perimeter is $2(2+6) = 16$ and $A = 2 \times 6 = 12$.

Now using the formula we have

$$T = ph + 2A = (16)(3) + 2(12) = 72 \text{ sq. ft. } Ans.$$

(b) Find the volume of a sphere of radius 6 inches.

Solution.

By the formula: $V = \frac{4}{3}\pi r^3 = \frac{4}{3}(3.1416)(6^3)$
$$= (288)(3.1416) = 904.7808 \text{ cu. in. } Ans.$$

(c) Find the total area and the volume of a right cone whose base is a circle of radius 6, and whose altitude is 8.

Solution.

The area of the base is

$$A = \pi r^2 = 36\pi.$$

The perimeter is $p = 2\pi r = 12\pi$.

The slant height is the hypotenuse of a right triangle with sides $r = 6$ and $h = 8$ so that

$$s = \sqrt{36 + 64} = \sqrt{100} = 10.$$

Now using the formulas we have

$$T = \tfrac{1}{2}ps + A = \tfrac{1}{2}(12\pi)10 + 36\pi$$
$$= 60\pi + 36\pi = 96\pi$$
$$= 301.5936. \ Ans.$$
$$V = \tfrac{1}{3}Ah = \tfrac{1}{3}(36\pi)(8) = 96\pi$$
$$= 301.5936. \ Ans.$$

Problems XXIII.

1. Find the volumes of the following solids:
   (a) A cube of side 6 inches.
   (b) A right cone of radius 3 and altitude 6.
   (c) A prism whose base is a rectangle 3 by 8 and whose altitude is 4.
   (d) A sphere of radius 3.8.
   (e) A right circular cylinder of radius 9 and altitude 2.
2. Find the total area of the following solids:
   (a) A spherical ball of radius 2 inches.
   (b) A right cylinder of radius 3.6 and altitude 2.8.
   (c) A frustum of a cone whose lower base is a circle of radius 7, upper base is a circle of radius 4, and slant height is 6.

3. What is the longest fishing pole that can be placed in a locker measuring 6' by 2' by 2'? (Hint: It will be the length of the diagonal between opposite corners.)

4. How many square feet of glass is needed to line a cylindrical water tank measuring 6 feet high with a diameter of 2 feet?

Review Exercises III.

1. Perform the indicated operations.
   (a) 29° 38' 24" − 20° 42' 36".   (b) (12° 10' 16")18.
   (c) Bisect 33° 18' 28".

2. Given the right triangle $a = 5$ and $c = 13$. Find the area.

3. Find the area of a trapezoid whose altitude is 5 inches if the parallel sides are 7 and 9 inches.

4. Find the trigonometric functions of 60°.

5. Find the circumference and area of a circle whose radius is 5.8 inches.

6. Find the area of a regular octagon whose sides are 11.6 inches.

7. Find the volume of a rectangular box which measures 3' 6" by 8' 9" by 10' 3".

8. How many square feet of wallpaper are needed to paper a room which is 16' by 12' and 8' high if we assume that the windows and doors take away about twenty per cent of the total area?

9. Find the volume of a cylindrical tank whose inside dimensions are a radius of 3 ft. 6 in. and a height of 6 ft. 6 in.

10. Show that the area of the shaded flower beds formed by describing semicircles on the three sides of the right triangle *ABC* of Figure 34 is equal to the area of the triangle. (These are known as the lunes of Hippocrates.)

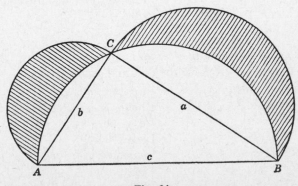

Fig. 34

Chapter V

WEIGHTS AND MEASURES

Introduction

As soon as man began to measure he was faced with the problem of establishing units of measure. Many of the earlier units came from man's experiences and parts of the body. There were such units as a day's journey; the span or the spread of the hand; the foot as the length of a man's foot. Measurements made in terms of these units caused considerable confusion since there were no standards and the length of a man's foot varied considerably from man to man. Surprisingly enough the confusion lasted for many years, and the first uniform system was not established until 1799 when France adopted the *metric* system. This system is used throughout most of the civilized world except the United States and Great Britain. The standard of length for the metric system is the meter, which is defined to be the distance between two marks on a platinum-iridium bar at the International Bureau of Weights and Measures near Paris. The definition further specifies that the distance should be measured when the bar is at 0° centigrade. The only legal relationship between the metric system and the one used in the United States is the equivalence

$$1 \text{ meter} = 39.37 \text{ inches}$$

which was established by an act of Congress in July, 1866. We shall discuss both the metric and the English system.

Measure of Time

Fortunately the measure of time is universal. The basic unit is the second, from which we have the following equivalences:

60 seconds (sec.)	= 1 minute (min.)
60 minutes	= 1 hour (hr.)
24 hours	= 1 day (da.)
7 days	= 1 week (wk.)
365 days	= 1 common year (yr.)
366 days	= 1 leap year
1 lunar month (mo.)	= 29 days, 12 hours, 44 minutes

The English System

I. Measures of Length. In the English system, distance is measured in the following units of length.

$$12 \text{ inches (in.)} = 1 \text{ foot (ft.)}$$
$$3 \text{ feet} \quad \text{(ft.)} = 1 \text{ yard (yd.)}$$
$$16\tfrac{1}{2} \text{ feet} \quad \text{(ft.)} = 1 \text{ rod (rd.)}$$
$$320 \text{ rods} = 1760 \text{ yards} = 5280 \text{ feet} = 1 \text{ mile.*}$$

II. Measures of Area. The surface measure is obtained in "square units" corresponding to the units of length. The common notation is, for example, sq. in. or in.². The following table gives equivalents in the English system.

$$144 \text{ square inches (in.}^2) = 1 \text{ square foot (sq. ft.)}$$
$$9 \text{ square feet} \quad \text{(ft.}^2) = 1 \text{ square yard (sq. yd.)}$$
$$30\tfrac{1}{4} \text{ square yards} \quad \text{(yd.}^2) = 1 \text{ square rod (sq. rd.)}$$
$$160 \text{ square rods} \quad \text{(rd.}^2) = 1 \text{ acre (A.)}$$
$$640 \text{ acres} \qquad\qquad = 1 \text{ square mile (sq. mi.)}$$

III. Measures of Volume. The amount of space that an object occupies is called the **volume**. Volume is also associated with the *capacity* of containers. The dimensions are in terms of cubic units such as the cubic inch or cubic foot. In the English system we have the following table of equivalence.

$$1728 \text{ cubic inches (cu. in.)} = 1 \text{ cubic foot (cu. ft.)}$$
$$27 \text{ cubic feet} \qquad\quad = 1 \text{ cubic yard (cu. yd.)}$$
$$128 \text{ cubic feet} \qquad\quad = 1 \text{ cord (cd.)}$$

In speaking of capacity in the United States † we adopt measures which are associated with containers, and there is a distinction between liquid measure and dry measure.

LIQUID MEASURES

$$4 \text{ gills (gi.)} = 1 \text{ pint (pt.)} = 16 \text{ fluid ounces}$$
$$2 \text{ pints} \quad\;\; = 1 \text{ quart (qt.)}$$
$$4 \text{ quarts} \;\; = 1 \text{ gallon (gal.)} = 231 \text{ cubic inches}$$
$$31\tfrac{1}{2} \text{ gals.} \quad = 1 \text{ barrel (bbl.)}$$

---

* There is also a **nautical mile**:
  1 nautical mile = 6076.10333 ft. = 1.15078 mi. = 1852 meters **(exactly)**.
† Some measures used in England are:
  1 British imperial quart (liquid) = 1.20095 U.S. liquid quarts.
  4 British imperial quarts (liquid) = 1 British imperial gallon.
  36 British imperial gallons = 1 British barrel.
  British dry quart = 1.0329 U.S. dry quarts.

DRY MEASURES

2 pints = 1 quart (qt.)
8 quarts = 1 peck (pk.)
4 pecks = 1 bushel (bu.) = 2150.42 cubic inches

IV. Measures of Weights. There are three kinds of weights in use in the United States. The **Avoirdupois** is the ordinary and is the one which is meant unless otherwise specified. The **Troy** weight is used by jewelers in weighing precious metals and stones. The **Apothecaries'** weight is used by druggists and physicians in compounding medicines, although the trend is for them to use the metric system.

The legal weight in the United States is really derived from the kilogram by an act of Congress on July 28, 1896, which established the relation 1 kilogram = 2.2046 pounds Avoirdupois. The National Bureau of Standards has revised this to be

1 kilogram   = **2.20462234 pounds Avoirdupois.**
1 pound Avoirdupois = 453.5924277 grams.

The relation between Avoirdupois and Troy is

1 Troy pound = $\frac{5760}{7000}$ Avoirdupois pound.

The following weights are in common use in the United States.*

AVOIRDUPOIS WEIGHT

16 drams (dr.) = 1 ounce (oz.)
7000 grains (gr.) = 1 pound (lb.)
16 ounces (oz.) = 1 pound
100 pounds   = 1 hundredweight (cwt.)
2000 pounds   = 1 ton (T.)
2240 pounds   = 1 long ton

TROY WEIGHT

24 grains   = 1 pennyweight (pwt.)
20 pennyweights = 1 ounce (oz.)
12 ounces   = 1 pound
5760 grains   = 1 pound
3.168 grains   = 1 carat

The Troy and Apothecaries' weights use the same pound and the same ounce. The Apothecaries' weights have the additional terms

* In England: 14 pounds = 1 stone; 112 pounds = 1 hundredweight.

20 grains = 1 scruple
3 scruples = 1 dram
8 drams   = 1 ounce

The term **karat** is a variation of carat and in this form is used in comparing the parts of gold alloys which are gold. Such comparison is based on the use of 24 karats to mean pure gold; therefore 14 karats means $\frac{14}{24}$ pure gold by weight or 14 parts pure gold and 10 parts alloy.

The definitions of weights and measures which have been listed above are used to change from one measure to another.

Examples

(a) Change the following measurements to inches.

(i) 6 feet. (ii) 2 yds. (iii) 3 yds. 2 ft. 11 in.

Solutions.

(i) 6 ft. $\times$ 12 = 72 in. *Ans.*
(ii) 2 yds. $\times$ 3 = 6 ft.; 6 ft. $\times$ 12 = 72 in. *Ans.*
(iii) 3 yd. $\times$ 3 $\times$ 12 = 108 in.
        2 ft. $\times$ 12 = 24 in.
        11 in. = 11 in.
—————————————————
3 yds. 2 ft. 11 in. = 143 in. *Ans.*

(b) Multiply 3 sq. ft. 120 sq. in. by 8.

Solution.

3 sq. ft. $\times$ 8   = 24 sq. ft.
120 sq. in. $\times$ 8 = 960 sq. in. =   6 sq. ft. 96 sq. in.
—————————————————————————
Total   = 30 sq. ft. 96 sq. in. *Ans.*

Problems XXIV

1. Change the following measurements to feet:
   (a) 19 yds. (b) 3 rods. (c) 2 miles 12 yds. (d) 3 yds. 6 ft. 9 in.
2. Find the number of seconds in 4 years 12 days and 35 minutes.
3. Find the number of square feet in 2 acres.
4. How many barrels are there in 4000 gallons?
5. How many pecks are there in 28 bushels?
6. Find the number of cubic inches in 3 cubic yards.
7. Find the number of ounces in 3 tons.
8. Change 8356 sq. in. to square yards, square feet, and square inches.
9. How many ounces (Troy weight) does a 3 carat diamond weigh?
10. If you waste 15 minutes a day for every day of the year, how much time would you waste in 2 years?

The Metric System

The simplicity of the metric system lies in the fact that it uses our decimal number system for equivalences. The system combines the basic unit with six numerical prefixes to coin very descriptive words.

PREFIX	ABBREVIATION	MEANING
milli	m	denotes 0.001
centi	c	denotes 0.01
deci	d	denotes 0.1
deka	dk	denotes 10
hecto	h	denotes 100
kilo	k	denotes 1000

The basic units are

> **meter** (m.), the unit of length;
> **liter** (l.), the unit of capacity;
> **gram** (g.), the unit of weight;
> **are** (a.),   the unit of area for land measure.

The measures are now easily obtained by combinations of the units with the prefixes. Thus for measures of *length* we have the equivalences:

10 millimeters (mm.)	= 1 centimeter (cm.)	= 0.01 meter
10 centimeters	= 1 decimeter (dm.)	= 0.1 meter
10 decimeters	= 1 meter (m.)	
10 meters	= 1 dekameter (dkm.)	= 10 meters
10 dekameters	= 1 hectometer (hm.)	= 100 meters
10 hectometers	= 1 kilometer (km.)	= 1000 meters

In measuring *surfaces* or *areas* we use the squares of the units of length. The most common ones are square centimeters (sq. cm. or cm.^{2}), square meters (sq. m. or m.^{2}), and square kilometers (sq. km. or km.^{2}). In land measure *the are is equal to 100 sq. m.*

The common measures of *volume* are the cubic centimeter (cc. or cm.^{3}), the cubic decimeter (dm.^{3}), and the cubic meter (m.^{3}).

The *measures of capacity* are based on the **liter,** which is defined as the volume of water at 4° C. and 760 mm. pressure which weighs 1 kilogram. In terms of cubic measure we have 1 liter = 1.000027 cubic decimeters, which is close enough so that for most practical purposes we say that 1 liter is 1 cubic decimeter. By combining the prefixes with the base liter we derive milliliters, centiliters, deciliters, etc. For practical purposes the kiloliter is 1 cubic meter and is frequently used.

The *measures of weight* are based on the **gram,** and the prefixes give the common units of weight. The gram is the weight of a cubic centimeter of water under the same conditions as defined for the liter. The most commonly used unit is the kilogram (kg.), and there is also a metric ton (tonneau) which is equal to 1000 kilograms.

For practical purposes we then have the very simple relations among the four units:

> 1 cubic decimeter equals 1 liter,
> 1 liter of water weighs 1 kilogram,
> 1 are is an area of 100 square meters.

Since the metric units employ 10 to a power in their relations, it is a relatively simple matter to change from one unit to another. Thus, for example, to change from centimeters to meters we divide by 100 or simply move the decimal point two places to the left.

Examples

(a) 4269 cm. = 42.69 m.
(b) 6834 g. = 6.834 kg.
(c) 8.39 hl. = 839 liters.

It should be remembered that in surface measure the scale is 100 and in volumes the scale is 1000.

Examples

(a) $5 \text{ m.}^2 = 500 \text{ dm.}^2 = 50,000 \text{ cm.}^2$
(b) $4 \text{ m.}^3 = 4000 \text{ dm.}^3 = 4,000,000 \text{ cm.}^3$

In recent years two new prefixes have been added to the language of weights and measures:

> **mega** — meaning one million
> **micro** — meaning one millionth

Examples

(a) 1 microgram = 0.000001 gram.
(b) 1 megagram = 1,000,000 grams.

Some of the prefixes have also found usage with the English units. Thus we often speak of kilotons and megatons, which mean 1000 tons and 1,000,000 tons, respectively.

Changing from One System to Another

The fundamental relations which permit us to change from the English system to the metric system or vice versa are shown in the following table:

1 m. = 39.37 in.	1 in. = 2.540005 cm.
1 m.² = 10.76387 sq. ft.	1 ft. = 30.48006 cm.
1 m.³ = 1.30794 cu. yd.	1 yd.² = 0.8361307 m.²
1 g. = 15.432356 grains.	1 yd.³ = 0.764559 m.³
1 kg. = 2.2046223 lbs.	1 lb. = 453.5924 g.

However, it is convenient to have more detailed equivalents for reference. These can be given in terms of conversion factors.

LENGTHS

To change from	to	Multiply by
inches	cm.	2.540005
feet	cm.	30.48006
yards	m.	0.914402
miles	km.	1.60935
centimeters	in.	0.3937
meters	ft.	3.28083
kilometers	mi.	0.621372

AREAS

To change from	to	Multiply by
sq. in.	sq. cm.	6.451626
sq. ft.	sq. m.	0.0929034
sq. yds.	sq. m.	0.8361307
acres	ares	40.46873
sq. mi.	sq. km.	2.589998
sq. cm.	sq. in.	0.155
sq. m.	sq. ft.	10.76387
sq. km.	sq. mi.	0.3861006

VOLUMES

To change from	to	Multiply by
cu. in.	cc.	16.38716
cu. ft.	cu. m.	0.028317
cu. yd.	cu. m.	0.764559
cc.	cu. in.	0.0610234
cu. m.	cu. ft.	35.3144
cu. m.	cu. yd.	1.30794

CAPACITY — LIQUID

To change from	to	Multiply by
ounces	cc.	29.57
pints	liters	0.473167
quarts	liters	0.946333
gallons	liters	3.785332
liters	fl. oz.	33.8147
liters	qts.	1.05671
liters	gal.	0.264178
cc.	ounces	0.033815

CAPACITY — DRY

To change from	to	Multiply by
pints	liters	0.550599
quarts	liters	1.101198
pecks	liters	8.80958
bushels	liters	35.2383
liters	pints	1.81620
liters	qts.	0.908102
dekaliters	pecks	1.13513
hectoliters	bu.	2.83782

WEIGHTS (AVOIRDUPOIS)

To change from	to	Multiply by
grains	grams	0.0647989
ounces	grams	28.349527
pounds	kg.	0.4535924
tons	kg.	907.18486
grams	grains	15.43235639
kilograms	lbs.	2.2046223
metric tons	lbs.	2204.6223

Examples

(a) Change the following measurements to centimeters:

    (i) 9 in.   (ii) 3 ft.   (iii) 2 ft. 3 in.

Solutions.

    (i) $9 \times 2.540005 = 22.860045$ cm.   *Ans.*

    (ii) $3 \times 30.48006 = 91.44018$ cm.   *Ans.*

    (iii) $2 \times 30.48006 = 60.96012$

        $3 \times 2.540005 = \underline{7.620015}$

                   68.580135 cm.   *Ans.*

(b) Change the following measurements to English units:

  (i) 4 km.   (ii) 3 kg.   (iii) 6 sq. m.   (iv) 11 liters.

Solutions.

  (i) $4 \times 0.621372 = 2.485488$ mi. *Ans.*
  (ii) $3 \times 2.2046223 = 6.6138669$ lbs. *Ans.*
  (iii) $6 \times 10.76387 = 64.58322$ sq. ft. *Ans.*
  (iv) $11 \times 0.264178 = 2.905958$ liquid gals. *Ans.*

Problems XXV.

1. Change 850 l. to liquid quarts; 38 m. to inches; 392 cm. to feet; 484 l. to dry quarts; and 12 kg. to pounds.
2. Find the number of cc. in 7 cu. in.
3. Change 112 sq. cm. to sq. in.
4. Find the number of metric tons in 12 English tons.
5. How many inches are there in 6 hectometers?

Applications

The units of measure form an important part of our everyday lives and, of course, are essential in all forms of business. In order to gain more experience with the terminology we have just learned, some typical problems will be discussed.

I. *Speed Problem.* If a car travels 60 miles per hour, what is the speed in feet per second?

Solution. To find the speed in feet per second we change from miles to feet and from hours to seconds. Since there are 5280 feet in a mile we have

$$60 \text{ mi.} = 60 \times 5280 \text{ ft.}$$

Since there are 60 seconds in 1 min. and 60 min. in 1 hr. we have

$$1 \text{ hr.} = 60 \times 60 = 3600 \text{ sec.}$$

Now the speed is given as a ratio; that is, miles per hour is mi.:hr. We can therefore form the proportion

$$\text{mi.:hr.} = \text{ft.:sec.}$$

or

$$\frac{60}{1} = \frac{60 \times 5280}{60 \times 60} = 88.$$

Therefore 60 mi./hr. = 88 ft./sec.

II. *The Produce Business.* A grocer buys potatoes by the bushel and sells them by the peck. If he bought 36 bushels at 75 cents a bushel and sold them at 29 cents a peck, how much profit did he make?

Solution. Since there are four pecks in one bushel, the grocer sold $(36)(4) = 144$ pecks at 29 cents or his sale was $(144)(0.29) = \$41.76$. He paid $(36)(0.75) = \$27.00$ for the potatoes. Thus his profit was

$$\$41.76 - \$27.00 = \$14.76. \quad Ans.$$

III. *Geometric Problems.* Floor carpeting is usually sold by the "yard," which means so much per square yard. However, room measurements are given in feet and inches; consequently, it is necessary to make a conversion.

Example

How much would it cost to carpet a 13′ 6″ by 11′ 9″ room with a carpet costing \$6.89 per square yard?

Solution. The area of the room is

$$(13'\ 6'')(11'\ 9'') = (13\tfrac{1}{2})(11\tfrac{3}{4}) = \tfrac{1269}{8} \text{ sq. ft.}$$

Since there are 9 sq. ft. in 1 sq. yd., we divide by 9.

$$(\tfrac{1269}{8})(\tfrac{1}{9}) = \tfrac{141}{8} = 17.625 \text{ sq. yds.}$$

The cost is

$$(17.625)(6.89) = \$121.44. \quad Ans.$$

This is the answer if the carpet roll can be cut to fit this room with no waste. However, this is usually not possible, and some allowance for waste must be made.

IV. *Kitchen Problems.* Measurements are important in cooking and baking even though we now have a considerable number of ready mixes. The commonly used measures in the kitchen are based on the kitchenware principle, but through common usage we have the following equivalences:

3 teaspoons = 1 tablespoon	$\tfrac{1}{2}$ cup = 1 gill
2 tablespoons = 1 oz. (liq.)	2 cups = 1 pint
16 tablespoons = 1 cup	4 cups = 1 quart

For solid foods, the following amounts usually weigh *one pound:*

2 cups of butter	$3\frac{3}{4}$ cups of flour
2 cups of shortening	$3\frac{1}{3}$ cups of flour (whole wheat)
$3\frac{1}{4}$ cups of corn meal	3 cups of raisins
$2\frac{2}{3}$ cups of dates (pitted)	$3\frac{1}{2}$ cups of sugar (brown)
$3\frac{1}{2}$ cups of dates (unpitted)	$2\frac{1}{4}$ cups of sugar (granulated)
$4\frac{1}{2}$ cups of grated cheese	$3\frac{3}{4}$ cups of sugar (powdered)

Food also measures differently depending upon the particular state that it is in. The following are good equivalences:

1 square of chocolate = 2 tablespoons of cocoa.
1 cup of broken uncooked macaroni = $2\frac{2}{3}$ cups cooked.
1 cup of uncooked rice = 4 cups cooked.
1 cup of broken uncooked spaghetti = 2 cups cooked.

Canned goods are packaged in various size cans:

No. 1 can = $1\frac{1}{2}$ cups	No. $2\frac{1}{2}$ can = $3\frac{1}{2}$ cups
No. 2 can = $2\frac{1}{2}$ cups	No. 3 can = 4 cups
No. 10 can = 13 cups	

Examples

(a) How much do $6\frac{1}{2}$ cups of sugar weigh?

Solution. Since $2\frac{1}{4}$ cups of sugar weigh one pound, 1 cup weighs $1 \div \frac{9}{4} = \frac{4}{9}$ pound. Then

$$6\frac{1}{2} \text{ cups weigh } (\tfrac{13}{2})(\tfrac{4}{9}) = \tfrac{26}{9} = 2\tfrac{8}{9} \text{ lbs.}$$

(b) How many liquid ounces are there in 8 tablespoons?

Solution. Since 2 tablespoons equal one ounce, one tablespoon equals $\frac{1}{2}$ ounce. Then

$$8 \text{ tablespoons} = 8(\tfrac{1}{2}) = 4 \text{ ounces.}$$

(c) What is the weight of 12 tablespoons of flour?

Solution. Since $3\frac{3}{4}$ cups weigh 1 pound, 1 cup weighs $\frac{4}{15}$ lb. Now

$$12 \text{ tablespoons} = \tfrac{12}{16} \text{ cups} = \tfrac{3}{4} \text{ cup.}$$

Therefore

$$12 \text{ tablespoons} = (\tfrac{3}{4})(\tfrac{4}{15}) = \tfrac{1}{5} \text{ pound.}$$

HANDY ANDY No. 12

Calorie Value Chart

Average Serving	Calorie Count	Average Serving	Calorie Count
Bread and Cereals		**Fruit Juices**	
white bread (1 slice)	64	grapefruit, fresh (1 cup)	87
whole wheat bread (1 slice)	55	orange, fresh (1 cup)	108
doughnut (1)	136	pineapple, canned (1 cup)	121
macaroni with cheese (1 cup)	464	tomato, canned (1 cup)	50
rice (1 cup)	201		
		Meat, Fish, and Poultry	
Beverages		beef, sirloin steak (3 oz.)	257
		lamb chop (3 oz.)	356
coffee or tea (plain)	0	pork chop (3 oz.)	284
carbonated beverages (8 oz.)	107	ham (3 oz.)	339
milk, whole (1 cup)	166	bacon (2 strips)	97
milk, skim (1 cup)	87	veal chop (3 oz.)	193
chocolate malted milk (2 cups)	562	chicken, canned, boned (3 oz.)	169
		tuna (3 oz.)	169
Dairy Foods			
butter (1 T.)	100	**Miscellaneous**	
cheese, cheddar (1 oz.)	113	mayonnaise (1 T.)	92
cheese, cottage (1 cup)	215	French dressing (1 T.)	59
cream, light (1 T.)	30	assorted jams (1 T.)	55
cream, whipped unsweetened (1 T.)	25	sugar (1 T.)	48
egg (1 medium-size)	77		
		Vegetables	
Desserts		beets (1 cup)	68
		broccoli (1 cup)	44
pie, apple (4 in. sector)	331	carrots (1 cup)	44
pie, custard (4 in. sector)	266	corn (1 cup)	140
pie, mince (4 in. sector)	341	peas (1 cup)	111
pie, lemon meringue (4 in. sector)	302	potatoes, mashed (1 cup)	159
cake, angel food (2 in. sector)	108	spinach (1 cup)	46
cake, layer (2 in. sector)	410	sweet potato, baked (medium-size)	183
brownie (3″ x 2″ x 2″)	295	tomato, raw (medium-size)	30
ice cream, plain (1/7 of a quart)	167		
sherbet (½ cup)	118	**"Just A Little Snack"**	
vanilla pudding (1 cup)	275	hamburger on a bun	492
		hot dog on a bun	300
Fruits		peanut butter sandwich	220
		cheese sandwich	333
apple, raw (medium-size)	76	chocolate nut sundae with	
banana, raw (medium-size)	88	whipped cream	350
cantaloupe (½)	37	cashews (1 oz.)	164
grapefruit (½ small)	49	peanuts, chopped (1 T.)	50
orange, raw (medium-size)	70	pecans, chopped (1 T.)	52
peach, raw (medium-size)	46	fudge (1 oz.)	116
pear, raw (medium-size)	95	chocolate creams (1 oz.)	110
pineapple, canned (1 large slice)	95		

Review Exercises IV.

1. Change the following measurements:
   (a) 3 miles 58 yards 9 feet to feet.
   (b) 3 acres to square feet.
   (c) 39 pecks to bushels.

2. Change the following measurements to English units:
    (a) 14 kilograms to pounds.
    (b) 19 liquid liters to quarts.
    (c) 134 cc. to cubic inches.
3. Change the following measurements to metric units:
    (a) 112 sq. yds. to square meters.
    (b) 9 cu. yards to cubic meters.
    (c) 6 quarts (dry measure) to liters.
4. If a gasoline distributor pays $7.50 a barrel and sells it for 31.9 cents a gallon, what is his profit on 1000 barrels?
5. A builder desires to pour a basement wall using readi-mix concrete which costs $12.75 a cubic yard. If the wall is to be 8 in. thick and 8 ft. high, what will it cost if the total length of all the walls is 180 ft.?

Chapter VI

GRAPHS

Charts and Graphs

It has long been known by teachers that visual aids are very important in the learning process. Charts and graphs are forms of visual aids; they show pictures of data and may be used to comprehend a situation quickly or to perform detailed analyses and computations. The two words "chart" and "graph" are often used interchangeably, and a clear distinction has not been recognized. Perhaps the best distinction is to limit the word *graph* to the picture which employs two or more scales and the word *chart* to all other types of visual presentations.

Charts and graphs are intended to show various things such as

(a) a comparison of the amounts of things;

(b) the rate of change of a thing;

(c) the relation of the parts to the whole;

(d) the relation of two or more quantities through a formula.

Charts are usually exhibited by pictorial means and frequently have some of the numerical data also written. Graphs, on the other hand, usually take the form of lines or curves on coordinate paper. We shall now discuss some typical charts and develop the construction of graphs.

Pictorial and Bar Charts

The bar chart and the 100% chart are the most frequently used charts in showing the relationship of the parts to the whole. Some examples are shown in Figure 35, p. 122. Such charts are fairly easy to read since much information is usually written on the chart.

To construct a circular chart, take the ratio of each part to the whole and multiply by 360°. The result is the number of degrees in the center angle of the sector to be formed by that part. Since the sum of the parts must equal the whole, the sum of the angles in the resulting sectors must total 360°.

Graphs

STOPPING DISTANCE FROM VARIOUS SPEEDS
(From 1958 Chart Developed by National Conference on Stopping Distances)

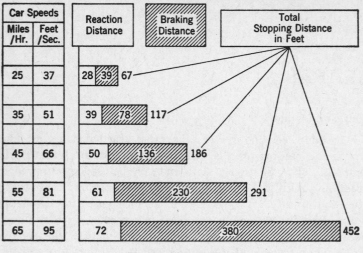

Car Speeds		Reaction Distance	Braking Distance	Total Stopping Distance in Feet
Miles /Hr.	Feet /Sec.			
25	37	28 39 67		
35	51	39 78 117		
45	66	50 136 186		
55	81	61 230 291		
65	95	72 380 452		

Fig. 35 (a) Bar Chart

INCREASE IN WORKING WOMEN
1955-1965
DISTRIBUTION BY AGE GROUPS

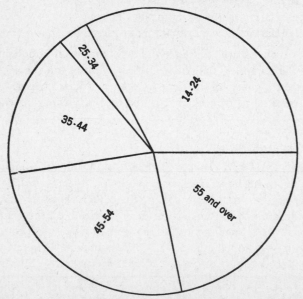

Fig. 35 (b) Circular 100% Chart

Example

The financial books of a corporation show the following division of expenditures:

$46\frac{1}{2}\%$ to supplies; 31% to employees; 11% for taxes; $3\frac{1}{2}\%$ for depreciation; 5% to stockholders; 3% for business.

Construct a circular graph to show the placement of the corporation's income.

Solution.

First obtain the central angles:

Supplies:	(.465)(360) =	167.4°
Wages:	(.31)(360) =	111.6°
Taxes:	(.11)(360) =	39.6°
Depreciation:	(.035)(360) =	12.6°
Stockholders:	(.05)(360) =	18.0°
Business:	(.03)(360) =	10.8°
Check:	Total =	360.0°

Draw a circle and a radius. Now using a protractor mark off the above central angles, drawing a radius each time. Label the figure (see Figure 36).

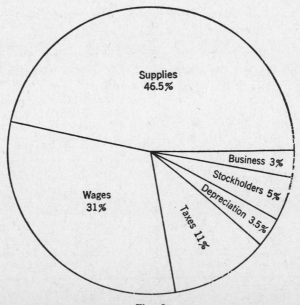

Fig. 36

To construct a bar chart, pick a scale either vertical or horizontal to represent the variable of comparison. Then draw bars to the units given in the data.

Example

Construct a bar graph to show the causes of accidents if 60% are caused by falls, 20% by fires, 5% by blows, 8% by automobile, and 7% by all others.

Solution. See Fig. 37.

ACCIDENTS (Causes)

60%   Falls

20%   Fires

5%   Blows

8%   Automobiles

7%   All Others

Fig. 37

Pictorial charts (frequently in the form of bar charts) are often used for dramatic effects. See Figure 38.

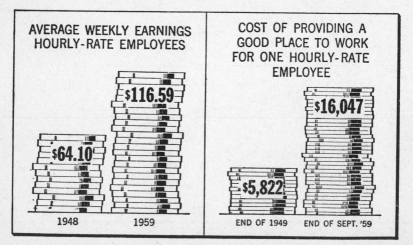

Fig. 38

Graphs

The main difference between graphs and charts is the fact that for graphs two or more scales are used. Suppose we draw two intersecting lines perpendicular to each other. We have now divided the plane of the paper into four *quadrants* which are usually numbered in a counterclockwise direction as shown in Figure 39.

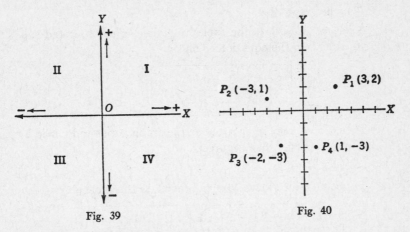

Fig. 39   Fig. 40

The point of intersection is called the **origin**, (*O*), and the two perpendicular lines are called the **axes**, *X* and *Y*. We shall now choose positive quantities to be measured upward from the horizontal line and to the right of the vertical line. Negative quantities are measured downward from the horizontal line and to the left of the vertical line. By choosing a scale on each of the axes we can locate any point if we know the x and the y value of its coordinates. If x is the horizontal value and y is the vertical value, then the following table is useful for remembering the sign in each quadrant.

Quadrant	I	II	III	IV
x-value	+	−	−	+
y-value	+	+	−	−

To **plot** a point, $P(x,y)$, is to locate its position from the values of x and y. In Figure 40 we have plotted the points $P_1(3,2)$, $P_2(-3,1)$, $P_3(-2,-3)$, $P_4(1,-3)$.

Cross-ruled papers of various kinds have been printed and can be purchased from book and school suppliers. When using such graph paper, first pick the axes and then the units to be used for each axis. We can now plot a formula which relates two variables. Such a graph may be made by first obtaining a table of values by giving values to one of the variables and calculating the corresponding values of the other variable from the formula.

Examples

(a) Plot the graph of $y = 2x - 3$.

Solution. Calculate the table by giving values to x and substituting into the equation to find y.

x	-1	0	1	2	3
y	-5	-3	-1	1	3

These points are then plotted on the graph paper and joined by a smooth curve. See Figure 41.

(b) Plot the graph of $y = x^2 + x - 6$.

Solution. Give values to x and calculate the table.

x	-4	-3	-2	-1	0	1	2	3
y	6	0	-4	-6	-6	-4	0	6

These points are plotted and joined by a smooth curve. The graph is shown in Figure 42.

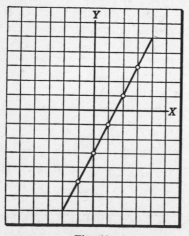

Fig. 41

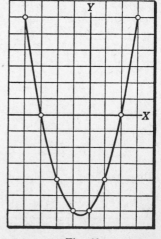

Fig. 42

Different Units on the Axes

For certain business graphs it is necessary to have different units for each axis, and frequently more than one curve is put on the same set of axes.

Example

A business concern has tabulated its sales and the cost of its overhead in thousands of dollars for a year as follows:

	Jan.	Feb.	Mar.	Apr.	May	June	July	Aug.	Sept.	Oct.	Nov.	Dec.
Sales	15	12	10	8	12	20	23	21	18	20	22	20
Overhead	2.2	2	2	2	2.1	2.2	2.5	2.2	2	2	2.2	2.2

Plot the graph.

Solution. Let the horizontal scale represent the months of the year and the vertical scale be $2000 per unit. The points are then plotted and joined by broken straight lines. The result is shown in Figure 43.

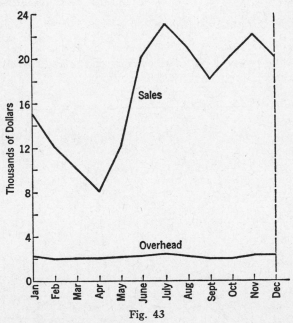

Fig. 43

Frequency Distribution Graphs

In statistics there are two problems which occur often. One is the amount of something at given times and the other is the amount of something over a given interval. The first of these is usually pictured by a broken line graph and the second by a bar graph often referred to as a *histogram*.

Examples

(a) The number of cars passing over a road counter is tabulated by the hour for a 24 hour period.

AM				PM			
Time	No.	Time	No.	Time	No.	Time	No.
12–1	90	6–7	175	12–1	160	6–7	75
1–2	75	7–8	380	1–2	75	7–8	25
2–3	15	8–9	425	2–3	25	8–9	115
3–4	10	9–10	275	3–4	215	9–10	100
4–5	40	10–11	110	4–5	475	10–11	75
5–6	105	11–12	130	5–6	300	11–12	80

Plot the data.

Solution. Let the X-axis represent the time and the Y-axis the number of cars. Plot the points and join them with broken lines. See Figure 44.

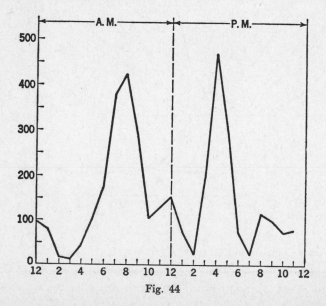

Fig. 44

(b) A doctor tabulated the weights of 200 of his adult male patients and then made a frequency table for a 5 pound interval.

Weights	No. of People	Weights	No. of People
150–154.9	3	180–184.9	23
155–159.9	7	185–189.9	21
160–164.9	15	190–194.9	15
165–169.9	20	195–199.9	9
170–174.9	45	200–204.9	5
175–179.9	34	205–209.9	3

Make a histogram of this table.

Solution. Let the horizontal axis represent the weights and pick a small square for each of the intervals. Let the Y-axis

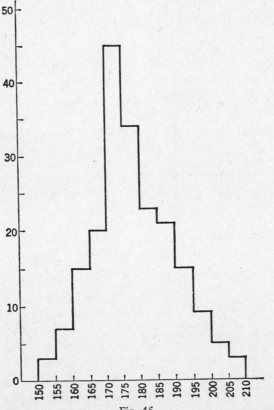

Fig. 45

represent the number of people. A histogram is a bar graph. The result is shown in Figure 45, p. 129.

Graphical Solutions of Equations

We studied the solution of simple equations in Chapter II. Equations may also be solved by graphical means. Consider for example the quadratic equation

$$x^2 + x - 2 = 0.$$

If we let the left-hand side of this equation be set equal to y we have a relationship between x and y which we can plot on graph paper,

$$y = x^2 + x - 2.$$

First calculate the table of values:

x	-3	-2	-1	0	1	2
y	4	0	-2	-2	0	4

and then plot these points as shown in Figure 46.

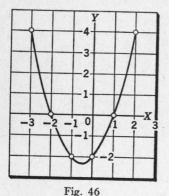

Fig. 46

Now the solutions of the original equation are those values for which y equals zero; in other words, the points where the curve crosses the X-axis. In this example, when $x = -2$ and $x = 1$. Let us consider a more difficult example.

Example

Find the solutions of

$$2x^3 - 5x^2 - 9x + 18 = 0$$

by graphical means.

Solution.

Let $y = 2x^3 - 5x^2 - 9x + 18$ and calculate.

x	$-2\frac{1}{2}$	-2	-1	0	1	2	3	$3\frac{1}{2}$
y	-22	0	20	18	6	-4	0	11

Plot these points and join them with a smooth curve. See Figure 47. The points at which the curve crosses the X-axis are -2, 1.5, 3; therefore, the solutions are $x = -2$, $x = 1.5$, $x = 3$. Notice that we have used different scales for each axis, which is permissible as we are only interested in the points of intersection with the *X*-axis.

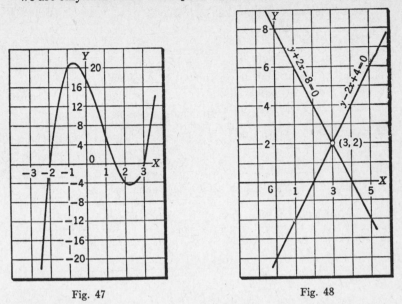

Fig. 47   Fig. 48

A linear expression involving *x* and *y* will plot as a straight line on a set of *X* and *Y* axes. If we now consider two such expressions at the same time we will have two lines which may intersect at a point. The coordinates of this point will satisfy the two equations and are said to be the solution to two *simultaneous linear equations*.

Example

Find the common solution to

$$y - 2x + 4 = 0,$$
$$y + 2x - 8 = 0.$$

Solution.

Calculate a table of values for each equation; since each is represented by a straight line only two points are needed; however, it is a good idea to get three points.

x	0	1	2
y	-4	-2	0

x	0	1	2
y	8	6	4

Plot the points and draw the lines. See Figure 48. The two lines intersect at (3,2). The solution to the problem is $x = 3$, $y = 2$.

Review Exercises V.

1. Obtain the necessary data and draw a graph of the following curves on an X, Y coordinate system:
   (a) $y = 3x + 2$.
   (b) $y = 2x^2 - 3x - 4$.
   (c) $y = x^3 - 7x - 6$.

2. Construct a circular graph to show the following data:
   36% red, 14% blue, 20% green, 15% orange, and 15% white.

3. Construct a bar graph for the data:

Ages	5–10	11–15	16–20	21–25	26–30	31–35	36–40	41–45	46–50
No.	5	9	13	36	35	21	18	7	2

4. Construct a graph for the following business data:

Month	Jan.	Feb.	Mar.	Apr.	May	June	July	Aug.
Sales ($1000)	9	11	8	10	15	20	25	25
Overhead ($1000)	3	3	4	5	5	6	6	6

5. Solve the simultaneous linear equations:
$$2x + 3y = 5,$$
$$x - 2y = 3.$$

6. The cost of the same make automobile has changed in the following manner:

Year	1940	1948	1952	1957	1960
Cost $	1250	1850	3050	3850	4250

Construct a graph to show these data.

7. Construct a graph to show the relationship between the area and radius of a circle.

8. Using the numbers 1 and all those divisible by 5, draw a graph showing the reciprocals of the numbers between 1 and 100. Using the graph find the approximate value of the reciprocal of 27.

9. Since 1 kilogram is approximately 2.205 pounds, draw a graph showing this relationship in the range 1 to 100 kilograms. From the graph determine the number of pounds in 38 kilograms.

10. The simple interest formula of $100 invested at 4% can give the amount in the form $A = 100 + 4t$ where t is the number of years. Draw a graph and determine in how many years the amount will be $160.

Chapter VII

LOGARITHMS

Definitions

Ever since man learned to add, subtract, multiply, and divide, he has attempted to improve his methods of doing arithmetic. The greatest improvements have come in recent years with the invention of the electronic calculators. However, the use of calculators goes back some 5000 years to the Japanese *abacus*, which is still in use today. For many years one of the best tools for doing arithmetic was the use of logarithms, and it is still an excellent tool since all the user needs is a table of logarithms.

Definition. *The* **logarithm** *of a number* **N** *to the base* **a** *is the exponent* **x** *of the power to which the base must be raised to equal the number* **N**.

For our discussion we shall limit ourselves to the common logarithms, which have the base equal to 10 and employ the notation

$$\log N \equiv \text{the logarithm of } N \text{ to the base 10.}$$

Then the definition states:

$$If \ n = 10^x, \ then \ x = \log N.$$

Let us write a table of numbers which are integral powers of 10 and the corresponding common logarithms.

Exponential Form	Logarithmic Form
. . . . .	. . . . .
$10^3 = 1000$	$\log 1000 = 3$
$10^2 = 100$	$\log 100 = 2$
$10^1 = 10$	$\log 10 = 1$
$10^0 = 1$	$\log 1 = 0$
$10^{-1} = 0.1$	$\log 0.1 = -1$
$10^{-2} = 0.01$	$\log 0.01 = -2$
$10^{-3} = 0.001$	$\log 0.001 = -3$
. . . . .	. . . . .

As long as we are dealing with numbers which are integral powers of 10, the logarithms are easy to find. Suppose now we consider a number which does not fall into that class, say $N = 115$. Since 115 is between 100 and 1000, it is natural to suppose, from the above table, that log 115 is between 2 and 3 or, in other words, the logarithm is 2+ (a proper fraction). Since we can express the proper fraction in decimal form, we have

$$\log N = (\textbf{an integer}) + (0 \leq \textbf{decimal fraction} < 1).^*$$

The integral part is called the **characteristic**. The decimal fraction is called the **mantissa**. Thus the above statement says

$$\log N = \textbf{characteristic} + \textbf{mantissa}.$$

Since the mantissa may be nonrepeating infinite decimals, they are approximated to as many places as desired. These approximations have been tabulated in four-place, five-place, or higher-place tables which are called logarithmic tables. Thus the mantissas, or decimal parts, are found from tables, and the values in the tables are **always positive**.

The characteristic is determined according to the following two rules:

Rule 1. *If the number* **N** *is greater than* **1,** *the characteristic of its logarithm is* **one** *less than the number of digits to the left of the decimal point.*

Rule 2. *If the number* **N** *is less than* **1,** *the characteristic of its logarithm is* **negative;** *if the first digit which is not zero occurs in the* **k**th *decimal place, the characteristic is* $-$**k.**

Since the characteristic and mantissa are combined to give the complete logarithm,

$$\log N = \textbf{characteristic} + \textbf{mantissa}$$

and, further, since the mantissa is always positive, it is best to write a negative characteristic, $-k$, as $(\textbf{10} - k) - \textbf{10}$.

Examples

(a) Write the logarithms having given a mantissa .6857 and characteristics 2, 0, -1, and -2.

---

* The notation $0 \leq$ decimal fraction < 1 means a decimal fraction between 0 and 1 and it can include 0.

Solution.

Characteristic	Logarithm
2	2.6857
0	0.6857
−1	9.6857 − 10
−2	8.6857 − 10

(b) Find the characteristics of the logarithms of the numbers in the left column.

Solution.

Number	Characteristic
396.7	2
39.42	1
7.283	0
0.3845	9 − 10
0.0642	8 − 10
0.000094	5 − 10

Suppose now that we have the logarithm given and desire to find the corresponding number. This number is called the **antilogarithm**. The characteristic of the given logarithm determines the position of the decimal point in the antilogarithm and the mantissa determines the digits in the antilogarithm. To place the decimal point, use *in reverse order* the two rules for determining the characteristic of the logarithm of a number.

Examples

The digits of an antilogarithm are 6452. Place the decimal point if the characteristic is 1, 2, 8–10.

Solution.

Characteristic	Number
1	64.52
2	645.2
8 − 10	0.06452

Tables and Their Use

We shall now turn to the problem of finding the mantissa of a logarithm. For this purpose we shall use the Table of Common Logarithms in the back of this book (Table IV). This table is a four-place table, and, although it will not yield as accurate results as a table of more places, it will serve for many problems.* First we must understand that the numbers .06041, .6041, and 604100.00 are said to have the *same sequence of digits*. The mantissa of the logarithm for each of these numbers is the *same;* the characteristics are, of course, different.

I. To find the logarithm of a given number.

First we determine the characteristic and then look up the mantissa in the table. Let us explain it by an example.

Examples

(a) Find log 23.4.

Solution. By Rule 1 the characteristic is 1. To find the mantissa turn to the table and locate the first two digits (23) in the left column headed by "N." In the "23" row and in the column headed by "4" (the third digit of the number) find 3692. This number is the mantissa.

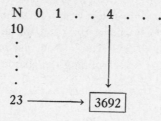

Thus log 23.4 = 1.3692.

(b) Find log 0.05736.

Solution. By Rule 2 the characteristic is 8 − 10. To find the mantissa we are faced with the problem of having more digits in the number than we have in the table and it will be necessary to *interpolate*. The procedure is to first locate "57" in the left column. In the "57" row and the column headed by "3" find

* For a five-place table see Kaj L. Nielsen, *Logarithmic and Trigonometric Tables* (New York: Barnes & Noble, Inc., 1961), pp. 2–21.

7582 and in the column headed by "4" find 7589. The difference 7589 − 7582 = 7 is now multiplied by .6 (6 being the fourth significant digit of the given number 0.05736): 7(.6) = 4.2, which is rounded off to the nearest whole number, 4. This number is added to the smaller mantissa: 7582 + 4 = 7586. The answer is given by

$$\log 0.05736 = 8.7586 - 10.$$

II. To find the antilogarithm of a given logarithm.

In finding the number which corresponds to a given logarithm we enter the table and perform the above steps in reverse order.

Examples

(a) Find N if $\log N = 7.6454 - 10.$

Solution. First find the mantissa 6454 in the table; it appears in the column headed by "2" in the row which has "44" at the left under "N." Thus the sequence of digits is 442. Now the characteristic is $7 - 10 = -3$. Using Rule 2 for the characteristic, the first digit after the decimal point should occur in the *third* place. The answer is $N = 0.00442.$

(b) Find N if $\log N = 1.5952.$

Solution. First seek the mantissa in the table. We find 5944 and 5955 corresponding to the numbers 393 and 394. Consequently, our number is between these two numbers. Now form the quotient

$$\left(\frac{\text{Given mantissa} - \text{smaller table mantissa}}{\text{Larger table mantissa} - \text{smaller table mantissa}}\right) 10$$
$$= \left(\frac{5952 - 5944}{5955 - 5944}\right) 10 = \frac{8}{11} (10) = \frac{80}{11} = 7+.$$

This number is rounded off to an integer which is "attached to" the *smaller* number N resulting from the table. Thus 7 is attached to 393 to give us the sequence of numbers 3937. The given characteristic is 1; so by Rule 1 we have

$$N = 39.37.$$

Problems XXVI.

1. Determine the characteristic for the logarithm of:
   (a) 3.25.   (b) 0.094.   (c) 356,000.   (d) 0.00031.

2. Find the logarithms of the following numbers:
   (a) 39.7.   (b) 0.3462.   (c) 839.4.   (d) 68.78.

3. Find the antilogarithms of the following logarithms:
   (a) 2.6952.   (b) 7.3465 − 10.   (c) 9.8638 − 10.

Using Logarithms for Computations

As we said at the beginning of this chapter, logarithms are used to ease our arithmetic. The use of logarithms for computation is based on the following four properties:

Property I. *The logarithm of a product is equal to the sum of the logarithms of its factors.*

$$\log MN = \log M + \log N.$$

Property II. *The logarithm of a quotient is equal to the logarithm of the numerator minus the logarithm of the denominator.*

$$\log \frac{M}{N} = \log M - \log N.$$

Property III. *The logarithm of the* **k**th *power of a number equals* **k** *times the logarithm of the number.*

$$\log N^k = k \log N.$$

Property IV. *The logarithm of the* **q**th *root of a number is equal to the logarithm of the number divided by* **q**.

$$\log \sqrt[q]{N} = \frac{1}{q} \log N.$$

These properties can be proved by the laws of exponents. Property IV gives us an easy method for finding *square roots* as then $q = 2$.

Example

Find $\sqrt{3556}$.

Solution. By Property IV

$$\log \sqrt{3556} = \tfrac{1}{2} \log 3556 = \tfrac{1}{2}[3.5509] = 1.7754.$$
$$\sqrt{3556} = \text{antilog } 1.7754 = 59.62. \quad Ans.$$

In carrying out computations using logarithms we should have a systematic form in which to put our work. This will eliminate many errors, and the following form is suggested. Check each step carefully.

Examples

(a) Find N if $N = \dfrac{5.367}{(12.93)(0.06321)}$.

Solution. We use logarithms and by the four properties of logarithms we write

$$\log N = \log 5.367 - [\log 12.93 + \log 0.06321].$$

(1) $\log 5.367 = 0.7298$	
(2) $\log 12.93 = 1.1116$	
(3) $\log 0.06321 = 8.8008 - 10$	
(4) $(2) + (3) = 9.9124 - 10$	
(5) $(1) - (4) = 0.8174$	$N = 6.567$

Note: To subtract (4) from (1) we first change 0.7298 to $10.7298 - 10$.

(b) Find N if $N = \dfrac{(\sqrt[3]{0.9573})(3.21)^2}{98.32}$.

Solution.

$$\log N = \tfrac{1}{3} \log 0.9573 + 2 \log 3.21 - \log 98.32.$$

(1)	$\tfrac{1}{3} \log 0.9573 = \tfrac{1}{3}[9.9810 - 10]$	$= 9.9937 - 10$
(2)	$2 \log 3.21 = 2[0.5065]$	$= 1.0130$
(3)	$(1) + (2)$	$= 11.0067 - 10$
(4)	$\log 98.32$	$= 1.9927$
(5)	$(3) - (4)$	$= 9.0140 - 10$
$N = 0.1033$		

Note: To find

$$\tfrac{1}{3}[9.9810 - 10] = \tfrac{1}{3}[29.9810 - 30] = 9.9937 - 10$$

always rearrange a negative characteristic so that after dividing the result will be x.xxxx minus an integer.

(c) Find N if $N = \sqrt{\dfrac{(0.3592)^3}{673.5}}$.

Solution.

	$\log N = \frac{1}{2}[3 \log 0.3592 - \log 673.5]$.	
(1)	$3 \log 0.3592 = 3[9.5553 - 10] = 28.6659 - 30$	
(2)	$\log 673.5$ $= 2.8284$	
(3)	$(1) - (2)$ $= 25.8375 - 30$	
(4)	$\frac{1}{2}(3)$ $= \frac{1}{2}[15.8375 - 20] = 7.9188 - 10$	

$$\boxed{N = 0.008294}$$

Note: In going from (3) to (4) we made the change

$$25.8375 - 30 = 15.8375 - 20.$$

This was done so that after dividing by 2 we would have x.xxxx − 10.

If the three sides (a, b, c) of a triangle are given, we can find the area of the triangle by the formula

$$A = \sqrt{s(s - a)(s - b)(s - c)}$$

where

$$s = \tfrac{1}{2}(a + b + c).$$

Example

Find the area of the triangle whose sides are $a = 139.4$, $b = 286.5$, $c = 242.7$.

Solution.

$a = 139.4$	$s - a = 194.9$	$\log (s - a) = 2.2898$
$b = 286.5$	$s - b = 47.8$	$\log (s - b) = 1.6794$
$c = 242.7$	$s - c = 91.6$	$\log (s - c) = 1.9619$
$2s = 668.6$		
$s = 334.3$	Sum $= 334.3$	$\log s = 2.5241$
	Check $= s$	$\log N = 8.4552$
		$\frac{1}{2} \log N = 4.2276$

Area $=$ antilog $\left[\frac{1}{2} \log N\right] = 16890.$

Note: A four-place table will give us only four significant digits in finding the antilog of a logarithm. Now if the characteristic requires more digits before the decimal point, zeros are annexed to the four significant digits.

Review Exercises VI.

1. Find the following numbers by the use of logarithms:

   (a) $N = \dfrac{(31.9)(123.4)}{(98.7)(23.56)}$.

   (b) $N = \dfrac{\sqrt{39.64}}{53.58}$.

   (c) $N = \dfrac{(\sqrt{3.49})3.142}{(1.83)^2}$.

2. A race track is built in the form of two equal straightaways of length 2342 feet each and two semicircles of radius 346.2 feet at each end. Find the distance around the track. [Hint: The mantissa for the log π is given to five places in the Table of Constants (Table II in the back of the book.)

3. Find the area of a right triangle given $a = 32.96$ and c (hypotenuse) $= 63.42$. [Hint: $\sqrt{c^2 - a^2} = \sqrt{(c-a)(c+a)}$.]

4. Find the area of a circle whose radius is 36.75 inches.

5. Find the volume of a circular cylinder if the radius of the base is 2.798 cm. and the height is 12.36 cm.

6. The rate of depreciation of equipment (such as your car) can be determined from the formula $W = C(1 - r)^n$, where W is the worth after n years, C is the original cost, and r is the rate of depreciation. Find the rate of depreciation of a car which is worth $525 after 5 years and originally cost $3670.

7. The period T (in seconds) of a pendulum of length l feet is given by

$$T = 2\pi \sqrt{\frac{l}{g}}.$$

Find T when $l = 2.25$ and $g = 32.16$.

Chapter VIII

MATHEMATICS OF GAMES

Introduction

Everyone is interested in games. Some games are games of skill, such as most athletic events; others are games of chance, such as dice or roulette; still others are combinations of skill and chance, such as most card games. In games of skill mathematics is used primarily to measure the performance of the players — for example, the batting averages of baseball players, the shooting averages of basketball players, bowling scores, golf scores, etc. In games of chance we can develop mathematical theories to explain the events, and we shall concern ourselves with these mathematical principles in this chapter.

Permutations and Combinations

Let us consider a set of objects which are distinguishable from each other; for example, 9 different people or the 11 different numbers resulting from the throw of a pair of dice. If we now wish to pick some of the objects from the set, we have to distinguish whether we want to pick them in a definite order or merely select them.

Definition. *Each different* **arrangement, a selection in a definite order,** *which can be made from a given number of objects, by taking any part or all of them at a time, is called a* **permutation.**

Fundamental Principle. If one thing can be done in p different ways, and after it has been done in any one of these ways, a second thing can be done in q different ways, then the two things can be done together, in the order stated, in pq different ways.

To illustrate this principle let us suppose that there are two ways to enter a building and three ways to exit from the building. Now how many ways can you enter and leave the building? According to the principle the answer is $(2)(3) = 6$. Let us identify the two entrances by A and B and the three exits by a, b, c. Then an entrance and exit can be accomplished in the following ways:

$$Aa, Ab, Ac, Ba, Bb, Bc.$$

The product of all the positive integers from **1** *to* **n**, *inclusive, is denoted by the symbol* **n**!:

$$1 \cdot 2 \cdot 3 \cdot 4 \cdot 5 \cdots n = n!$$

It is read "*n* factorial." The values of factorials for $1 \leq n \leq 9$ are given in the following table:

n	1	2	3	4	5	6	7	8	9
$n!$	1	2	6	24	120	720	5040	40320	362880

Let us now consider the computation of various permutations. The symbol $P(n,r)$ is used to denote *the number of permutations of* **n** *things taken* **r** *at a time*. We have the following formulas:

I. The permutation of *n* *different* things taken *r* at a time is

$$P(n,r) = n(n-1)(n-2) \cdots (n-r+1)$$
$$= \frac{n!}{(n-r)!}.$$

II. The permutation of *n* different things taken *n* at a time is

$$P(n,n) = n!$$

III. The number of distinct permutations of *n* things taken **all** at a time of which *p* are alike, *q* others are alike, *r* others are alike, etc. is

$$P = \frac{n!}{p!\,q!\,r!\cdots}.$$

Examples

(a) Find the values of $P(8,2)$, $P(21,3)$, $P(7,7)$.

Solution. Using formula I above:

$$P(8,2) = 8 \cdot 7 = 56.$$
$$P(21,3) = 21 \cdot 20 \cdot 19 = 7980.$$
$$P(7,7) = 7! = 5040.$$

(b) Eight swimmers compete in a race in which only the first three places receive medals. In how many ways can the medals be distributed?

Solution. This is a permutation of 8 things taken 3 at a time. Using formula I above:

$$P(8,3) = 8 \cdot 7 \cdot 6 = 336.$$

(c) How many permutations can be made of the letters in the word *Illinois* when taken all at a time?

Solution. This is a permutation of 8 things taking all of them at a time. However, some of them are alike; namely, $3i$ and $2l$. Thus using formula III above:

$$\frac{8!}{3!\,2!} = 8 \cdot 7 \cdot 6 \cdot 5 \cdot 2 = 3360.$$

Another method of choosing a set of objects is to make a selection without regard to order.

Definition. *All the possible* selections *consisting of* r *different things chosen from* n *given things* (r $\leq$ n), *without regard to the order of selection, are called* combinations.

Symbolically, we write $C(n,r)$ which is read "The number of combinations of n things taken r at a time." The formula for evaluation is

$$C(n,r) = \frac{P(n,r)}{r!} = \frac{n!}{r!(n-r)!}$$

Examples

(a) Find the values of $C(6,3)$, $C(28,25)$, and $C(12,7)$.

Solution.

$$C(6,3) = \frac{6!}{3!\,3!} = \frac{6 \cdot 5 \cdot 4}{3 \cdot 2 \cdot 1} = 20.$$

$$C(28,25) = \frac{28!}{25!\,3!} = \frac{28 \cdot 27 \cdot 26}{3 \cdot 2 \cdot 1} = 3276.$$

$$C(12,7) = \frac{12!}{7!\,5!} = \frac{12 \cdot 11 \cdot 10 \cdot 9 \cdot 8}{5 \cdot 4 \cdot 3 \cdot 2 \cdot 1} = 792.$$

(b) In how many ways can a hand of thirteen cards be dealt from a deck of fifty-two cards so as to contain exactly 10 spades, if the spades are selected first?

Solution. There are 13 spades in a normal deck from which we must select 10. This leaves 39 cards from which we must select 3. Thus

$$C(13,10) = \frac{13!}{10!\,3!} = 286.$$

$$C(39,3) = \frac{39 \cdot 38 \cdot 37}{3 \cdot 2 \cdot 1} = 9139.$$

By the fundamental principle, the number of ways this can be done together is the product

$$(286)(9139) = 2,613,754.$$

Elementary Probability

We shall now discuss some elementary concepts of probability which can be applied to games of chance. Suppose an event can result in one of n ways, each of which, in a single trial, is equally likely. Let s of these ways be considered successes and the others, $n - s$, failures. Then the probability, p, that one particular trial will result in a success is given by

$$p = \frac{s}{n}.$$

The probability, q, that one trial will result in a failure is given by

$$q = \frac{n - s}{n}.$$

It is now easy to see that

$$p + q = \frac{s}{n} + \frac{n - s}{n} = 1,$$
$$q = 1 - p, \ (0 \le p \le 1).$$

If p is the probability of success in one trial of an event, then pk is the **probable number** of success in k trials. If a person is to receive a sum of S dollars in case a certain event results successfully with a probability p, then the value of his **expectation** is pS dollars.

Examples

(a) What is the probability that one card drawn from a deck of 52 cards will be a spade? an ace? the ace of spades?

Solution. There are 52 cards; therefore $n = 52$.

  (i) There are 13 spades; thus $s = 13$ and $p = \frac{13}{52} = \frac{1}{4}$.

  (ii) There are 4 aces; thus $s = 4$ and $p = \frac{4}{52} = \frac{1}{13}$.

  (iii) There is one ace of spades; thus $s = 1$ and $p = \frac{1}{52}$.

(b) What is the probability of getting a head in the toss of a coin?

Solution. There are two possibilities in tossing a coin, either heads or tails. There is only one way in which heads can be accomplished. $n = 2$ and $s = 1$ so that $p = \frac{1}{2}$.

Applications to Games of Chance

Let us consider the game of dice played with two dice each having numbers from 1 to 6, inclusive. There are then six possibilities for each die so that the possibilities for 2 dice, by the fundamental principle, is $(6)(6) = 36$. Now if both turn up a "one" the resulting number is 2 and there is only one way in which this can occur. However, the number 3 can occur in two ways; namely, a 1 on die number 1 and a 2 on die number 2 or a 2 on die number 1 and a 1 on die number 2. The number 7 can be made with 1 and 6, 2 and 5, 3 and 4 and then with these numbers reversed so that 7 can be made in 6 ways. The complete list of possibilities and the corresponding probabilities are given in the following table:

Number	Possibilities	Probability
2	1	$\frac{1}{36}$
3	2	$\frac{1}{18}$
4	3	$\frac{1}{12}$
5	4	$\frac{1}{9}$
6	5	$\frac{5}{36}$
7	6	$\frac{1}{6}$
8	5	$\frac{5}{36}$
9	4	$\frac{1}{9}$
10	3	$\frac{1}{12}$
11	2	$\frac{1}{18}$
12	1	$\frac{1}{36}$

Thus the chance of "throwing a 7" is 1 in 6.

The same kind of reasoning can be applied to the probability of drawing certain cards in a card game. Suppose in a straight 5 card draw poker game with five players you are dealt four hearts and desire to draw one card to complete a heart flush. The fact that there are five players in the game does not enter into the mathematical probability. As far as you are concerned, there are 52 (the number of cards in the deck) − 5 (the number of cards in your hand) = 47 cards out. There are 13 (total number) − 4 (those in your hand) = 9 hearts out. Thus the probability is $\frac{9}{47}$ of drawing another heart. Remember, however, that the game of poker is not only a game of chance but also one of skill. The best way to enjoy a card game is not to rely on mathematical probabilities but to perfect your skill and play the luck of the particular evening of the game.

Review Exercises VII.

1. Find the values of
   (a) $P(9,3)$.   (b) $P(28,5)$.   (c) $P(15,4)$.
2. Find the values of
   (a) $C(9,3)$.   (b) $C(28,22)$.   (c) $C(15,4)$.
3. In presenting 3 men to 5 women, how many introductions are made?
4. How many integers, each of four different digits, can be formed from the digits 1, 2, 4, 5, 7, 9?
5. How many committees of 6 people can be selected from a group of 12 people?
6. How many permutations can be made of the letters in the word *trigonometry* when taken all at a time?
7. There is one winning ticket in a box containing 100 tickets and it pays $10.00. What is the value of the expectation of a single draw?
8. What is the probability of drawing a black card from a standard deck of 52 cards? What is the probability of drawing a black ace?
9. A man is to receive $10 if he throws a seven with a pair of dice. What is the value of his expectation?
10. What is the probability of drawing a card bigger than a jack from a deck of 52 cards?

Chapter IX

MATHEMATICS OF MONEY

Simple Interest

When money is borrowed and loaned, interest is charged for the use of the money. The amount of such interest depends upon the amount of the *principal* (amount borrowed), the length of *time* it is borrowed, and the *rate* of interest (usually expressed in per cent). If we let I = interest, P = principal, r = rate, and t = time (in years), then the *simple interest* is computed by the formula

$$I = Prt.$$

The *total amount*, A, is the principal plus the interest, thus

$$A = P + I.$$

Examples

(a) Find the simple interest on $500 for 2 years at 3%.

Solution.

$$I = (500)(\tfrac{3}{100})(2) = \$30. \quad Ans.$$

(b) Find the amount due on $1000 at the rate of $5\tfrac{1}{2}\%$ for 6 years.

Solution.

$$I = 1000(\tfrac{55}{1000})(6) = \$330.$$
$$A = P + I = \$1000 + \$330 = \$1330. \quad Ans.$$

Although the interest and time are quoted in per cent for so many years, money is frequently borrowed and loaned for shorter periods than one year.

When money is invested for a certain number of days, there are two ways to calculate the interest. If each day is to be $\tfrac{1}{365}$ of a year, the interest is called **exact interest.** If each day is to be $\tfrac{1}{360}$ of a year, the interest is called **ordinary interest.** Most business men use ordinary interest, while banks use exact interest. Common periods are 60 days and 90 days; then for ordinary interest we have $\tfrac{60}{360} = \tfrac{1}{6}$ and $\tfrac{90}{360} = \tfrac{1}{4}$.

Examples

(a) Mr. Jones needs $1200 to pay for some merchandise. If he borrowed the money at 6% for 90 days, how much ordinary interest would he pay? How much exact interest?

Solution. For ordinary interest

$$I = 1200(\tfrac{6}{100})(\tfrac{90}{360}) = 12(6)(\tfrac{1}{4}) = \$18. \quad Ans.$$

For exact interest

$$I = 1200(\tfrac{6}{100})(\tfrac{90}{365}) = 12(6)\tfrac{18}{73} = \$17.75. \quad Ans.$$

(b) Find the ordinary interest on $600 invested at 5% for 45 days.

Solution.

$$I = 600(\tfrac{5}{100})(\tfrac{45}{360}) = 6(5)(\tfrac{1}{8}) = \$3.75. \quad Ans.$$

Money is also borrowed by the month, and in such cases we use 12 months in the year and corresponding dates for the month. Thus a sum borrowed for 3 months on January 5 would be due on April 5.

Example

If you borrow $2400 to remodel a home at $5\tfrac{1}{2}$% and can repay it in 10 months, how much would you pay?

Solution.

$$A = P + I = 2400 + 2400(\tfrac{55}{1000})(\tfrac{10}{12})$$
$$= 2400 + 2(55) = 2510. \quad Ans.$$

Compound Interest

Money placed in a savings account with a bank is money that the depositor lends the bank, and interest is paid by the bank for the use of that money. Furthermore, the interest is periodically added to the principal and it in turn earns interest. This method of computing interest is called **compound interest**. The interest is converted to principal at the end of a certain time period such as each year, or half year, etc. These periods are called **conversion intervals**. The usual terminology is to say, "the interest is compounded annually, semi-annually, or quarterly." The total amount due is called the **compound amount** and is the principal plus the compound interest.

We can calculate the amount by calculating the interest at the end of each period and adding it to the previous amount.

Example

(a) Find the compound interest on $600 for 2 years at 4% per annum compounded semiannually.

Solution.

Original principal	$600.00
Interest on $600 at 4% for ½ yr. [600(.04)(.5)]	12.00
Amount at end of 1st period	$612.00
Interest on $612 at 4% for ½ yr.	12.24
Amount at end of 2nd period	$624.24
Interest on $624.24 at 4% for ½ yr.	12.48
Amount at end of 3rd period	$636.72
Interest on $636.72 at 4% for ½ yr.	12.73
Amount at end of 2 years	$649.45
Deduct original principal	600.00
Compound interest	$ 49.45 *Ans.*

(b) Compute the simple interest on $600 at 4% for 2 years and compare with example (a).

Solution.

$$I = Prt = (600)(.04)(2) = \$48.00. \quad Ans.$$

Compound interest − simple interest

$$= \$49.45 - \$48.00 = \$1.45. \quad Ans.$$

Computing compound interest in the manner of example (a) is far too tedious. Consequently, compound interest tables have been computed and are used by financial institutions. These tables are based on the formula

$$A_n = P(1 + r)^n$$

where A_n is the amount that a principal of P placed at a rate of interest r, compounded annually, for n years will accumulate. The table lists entries for a principal of $1; i.e., the table really gives the numbers for $(1 + r)^n$. To obtain the amount on P dollars we multiply the table entry by P.

Example

Find the compounded amount of $500 for 5 years at 5% compounded annually.

Solution. From Table V in the column headed by 5% and in the row marked 5 (under n) we find the number 1.2763. Therefore

$$A = 500(1.2763) = \$638.15. \quad Ans.$$

The same table can be used to find the amount for other conversion intervals than annually. For that purpose we need the fact that:

A principal **P** *placed at a rate of interest* **r** *and compounded* **q** *times a year for* **n** *years will grow to an amount* **A$_n$** *according to the formula*

$$A_n = P\left(1 + \frac{r}{q}\right)^{nq}.$$

Thus to use the table we divide the interest rate r by q, multiply the number of years n by q, and then enter the table with these new numbers.

Example

Find the compound interest on $600 for 2 years at 4% per annum compounded semiannually.

Solution. Semiannually means twice a year; thus $q = 2$. Calculate the new rate and period.

$$r = \frac{4\%}{2} = 2\%; \quad n = (2)(2) = 4.$$

Enter the table at $r = 2\%$ and $n = 4$ to obtain the number 1.0824. Then

$$A = 600(1.0824) = \$649.44. \quad Ans.$$

We worked this example the "long way" and in comparing the two answers we see that they differ by one cent. Such errors occur because of rounding off. To eliminate the errors financial institutions use tables which have seven or more significant digits.

Problems XXVII.

1. Find the simple interest on $800:
   (a) for 4 years at 4%;
   (b) for 3 years at 6%;
   (c) for 7 years at $2\frac{1}{2}$%;
   (d) for 45 days at 12% (ordinary);
   (e) for 73 days at 11% (exact);
   (f) for 7 months at $5\frac{1}{2}$%.
2. Find the compound amount on $1200 invested:
   (a) for 3 years at 4% compounded semiannually;
   (b) for 2 years at 6% compounded quarterly;
   (c) for 20 years at 3% compounded annually.
3. How long will it take $100 to double itself if invested at 4% compounded semiannually?

Borrowing Money

When an individual borrows money from a bank, a loan company, or a credit union, there are a number of ways to figure the interest and method of payment. A loan from a bank with proper collateral is frequently made on a promissory note. In such cases the interest is usually collected in advance and is called a **bank discount**. The amount of the loan is called the **face** of the note and is the amount that the borrower must return on the date that the note falls due, which is said to be at **maturity**. The difference between the face of the note and the discount is called the **net proceeds** and is the actual amount that the borrower receives at the time he gets the loan. The length of time that the money is borrowed may be expressed as so-many-months or so-many-days. In each case the interest is calculated for the number of days the money was used. The difference is in the calculation of the date of maturity. For a note by the month, the date of maturity agrees with the date that the loan was made.

DATE OF NOTE	TERM OF NOTE	DATE OF MATURITY
Feb. 1	3 months	May 1
June 10	2 months	Aug. 10
Sept. 30	1 month	Oct. 31
Nov. 30	3 months	Feb. 28

Remark: If the note is made on the last day of a month, it will fall due on the last day of the resulting month.

In calculating the maturity date for a note in which the terms are stated in days, the actual days are counted; usually the day on which the note is dated is not counted, but the date of payment is counted.

Examples

What is the date of maturity of a note dated on June 7 and for the term of (a) 60 days? (b) 90 days?

Solutions.

(a) June $(30 - 7) = 23$
     July $= 31$
       Total $= \overline{54}$
     Aug. $= 6$
       Total $= \overline{60}$

Therefore the date of maturity is Aug. 6. *Ans.*

HANDY ANDY No. 13											
Table of Days between Two Dates											

Mo. Day	Jan.	Feb.	March	April	May	June	July	Aug.	Sept.	Oct.	Nov.	Dec.
1	1	32	60	91	121	152	182	213	244	274	305	335
2	2	33	61	92	122	153	183	214	245	275	306	336
3	3	34	62	93	123	154	184	215	246	276	307	337
4	4	35	63	94	124	155	185	216	247	277	308	338
5	5	36	64	95	125	156	186	217	248	278	309	339
6	6	37	65	96	126	157	187	218	249	279	310	340
7	7	38	66	97	127	158	188	219	250	280	311	341
8	8	39	67	98	128	159	189	220	251	281	312	342
9	9	40	68	99	129	160	190	221	252	282	313	343
10	10	41	69	100	130	161	191	222	253	283	314	344
11	11	42	70	101	131	162	192	223	254	284	315	345
12	12	43	71	102	132	163	193	224	255	285	316	346
13	13	44	72	103	133	164	194	225	256	286	317	347
14	14	45	73	104	134	165	195	226	257	287	318	348
15	15	46	74	105	135	166	196	227	258	288	319	349
16	16	47	75	106	136	167	197	228	259	289	320	350
17	17	48	76	107	137	168	198	229	260	290	321	351
18	18	49	77	108	138	169	199	230	261	291	322	352
19	19	50	78	109	139	170	200	231	262	292	323	353
20	20	51	79	110	140	171	201	232	263	293	324	354
21	21	52	80	111	141	172	202	233	264	294	325	355
22	22	53	81	112	142	173	203	234	265	295	326	356
23	23	54	82	113	143	174	204	235	266	296	327	357
24	24	55	83	114	144	175	205	236	267	297	328	358
25	25	56	84	115	145	176	206	237	268	298	329	359
26	26	57	85	116	146	177	207	238	269	299	330	360
27	27	58	86	117	147	178	208	239	270	300	331	361
28	28	59	87	118	148	179	209	240	271	301	332	362
29	29	...	88	119	149	180	210	241	272	302	333	363
30	30	...	89	120	150	181	211	242	273	303	334	364
31	31	...	90	...	151	...	212	243	...	304	...	365

For Leap Year, One Day Must Be Added to Each
Number of Days after February 28.

(b) June (30 − 7) = 23
   July   = 31
   Aug.   = 31
    Total   = 85
   Sept.   = 5
    Total   = 90

Therefore date of maturity is Sept. 5. *Ans.*

To find the amount of the discount and net proceeds, ordinary interest (year = 360 days) is usually used and the interest is calculated from the date of the discount to the date of maturity.

Examples

(a) A 60-day note for $600 is dated April 5 and is discounted on that day at 7%. Find the discount and net proceeds.

Solution.

$$\$600(\tfrac{7}{100})(\tfrac{60}{360}) = \$7.00, \text{ discount. } Ans.$$
Net proceeds = $600 − $7 = $593. *Ans.*

(b) A bank note for 3 months for $600 was dated and discounted on April 5 at 6%. Find the amount of the discount and net proceeds.

Solution.

Date of maturity is July 5 and the term of discount is

    April (30 − 5) = 25
    May   = 31
    June   = 30
    July   = 5
          91 days

$$600(\tfrac{6}{100})(\tfrac{91}{360}) = \$9.10, \text{ discount. } Ans.$$
Net proceeds = $600.00 − $9.10 = $590.90. *Ans.*

Sometimes loans are made which may be repaid in installments. These installments may be in equal amounts or may be a certain principal plus the interest for a specified time (some credit unions which charge 1% per month on the unpaid balance use this method). Installment buying is a form of this type of borrowing money.

If all the installments are equal, a formula which approximates the true interest rate very well is given by

$$r = \frac{2NI}{B(n+1)}$$

where

r is the interest rate expressed as a decimal;

N is the number of payments in one year (for monthly payments $N = 12$; weekly payments $N = 52$);

I is the total interest in dollars;

B is the unpaid balance at the beginning of the credit period;

n is the number of installment payments.

Example

A loan company lends \$125 for 4 months with equal monthly install-ment payments of \$33.00. What is the annual rate?

Solution.   Amount paid 4(\$33)   = \$132
                  Amount of loan   = $\underline{\;125 = B}$
                Total interest paid = $7 = I$

Now $N = 12$ and $n = 4$, so that we have

$$r = \frac{2NI}{B(n+1)} = \frac{2(12)(7)}{125(5)} = \frac{168}{625} = 26.9\%.$$

Problems XXVIII.

1. Find the date of maturity of a note dated July 5, 1960, and for terms of
   (a) 30 days.   (b) 45 days.   (c) 90 days.
   (d) 2 months.   (e) 6 months.   (f) 2 years 3 months.
2. Find the discount and net proceeds of a 90-day note for \$500 dated on July 1 and discounted on that date for $6\frac{1}{2}\%$.
3. A bank makes a loan of \$300 for 15 months on a plan which requires monthly payments of \$21.50 per month. What is the annual rate of interest?

Present Worth

We have seen that money invested at some interest rate will grow in value as the years pass; for example, \$100 placed in a savings account which pays 3% compounded semiannually will be worth \$119.56 in six years. Thus the *future* value of a *present* amount of money is greater than its *present* value. It is therefore reasonable that the *present worth* of a sum of money which is due later is less than its future value. The present worth can be calculated in a manner similar to that for obtaining the compound amount. What we do is to find the value of P in the formula

$$A_n = P(1 + r)^n$$

when A_n, r, and n are given. Thus

$$P = A_n(1 + r)^{-n}.$$

To speed the calculations we can use a table of numbers for the present worth of \$1 compounded annually. (See Table VI.) If the conversion interval is not annually, then we can use this table by dividing the interest rate r by q and multiplying the number of years n by q, where q is the number of times a year that the amount is compounded.

Examples

(a) Find the present worth of \$1000 due in 5 years with interest at 3% compounded annually.

Solution. In this problem we have $n = 5$ and $r = 3\%$. From Table VI we have .86261 so that

$$P = \$1000(.86261) = \$862.61. \quad Ans.$$

(b) Find the present worth of \$500 due in 4 years with interest at 3% compounded semiannually.

Solution. In this problem we have $r = 3\%$, $n = 4$, and $q = 2$.

Thus we change the rate to $\dfrac{r}{q} = 1.5\%$ and the time to $nq = 8$.

We now enter the table at 1.5% and 8 to obtain .88771. Then

$$P = \$500(.88771) = \$443.86. \quad Ans.$$

Annuities

The subject of annuities and its allied phases is an important part of modern business as well as a matter of concern to most individuals. There is hardly a business, whether it be big or small, that is not saddled with some financial obligation which will have to be met at a future date. By setting aside small sums of money at fixed intervals of time and investing them at compound interest, the final value of these invested amounts can be made to equal the obligation at a future date. This is the purpose of a **sinking fund**. This method is also used to assure an income after retirement, to cancel a mortgage, or to provide a college education in the future. Annuities date back as far as the ancient Romans.

An **annuity** *is a sum of money to be paid at fixed intervals of time.*
There are various types of annuities. An **annuity certain** is an annuity
that begins at a specified time and terminates at a specified time.
The **contingent annuity** is an annuity that depends upon some event,
such as the death of the person concerned.

The two major quantities involved in problems on annuities are
the amount of the annuity (how much will you have at the end?) and
the present value of an annuity (how much is the end amount worth
today?). The total sum of payments and interest due at the end of
the term of an annuity is called the **amount** of the annuity, in other
words, the money which will be available at the end of a given time.
The **present value** of an annuity is the sum of the present values of all
the payments of the annuity; in other words, the cash equivalent of
the annuity at the present time (today).

Problem 1. *To find the amount of an annuity of* S *dollars per year,*
payable annually for **n** *years.* Let A = the amount and i = effective
interest rate; then

$$A = \frac{S}{i}\left[(1 + i)^n - 1\right].$$

If the interest is converted m times a year at a yearly interest rate of j,
then

$$A = \frac{S\left[(1 + j/m)^{mn} - 1\right]}{(1 + j/m)^m - 1}.$$

To evaluate these formulas requires a fair amount of arithmetic and
time. Financial institutions have computed and tabulated elaborate
annuity tables in order to cut down on the amount of work. The
cumbersome part of the calculations is the evaluation of the terms
$(1 + i)^n$ and $(1 + j/m)^{mn}$. This can be done by logarithms, but to
get the desired accuracy, we need to use at least a five-place logarithm
table. Since these two quantities are usually between the values of
1.000 and 1.100, we can use an abbreviated table providing n is not
too large. We present the abbreviated table for 100–150 on p. 158.*

Examples

(a) A man puts $300 into a financial institution at the end of every
year for 5 years. Find the amount of his savings at the end of
5 years if the institution pays 3%, payable annually.

* For a complete table see Kaj L. Nielsen, *Logarithmic and Trigonometric*
Tables (New York: Barnes & Noble, 1961). pp. 2–21.

Mathematics of Money

Common Logarithms for 100–150

N	L O	1	2	3	4	5	6	7	8	9
100	00 000	043	087	130	173	217	260	303	346	389
101	432	475	518	561	604	647	689	732	775	817
102	860	903	945	988	*030	*072	*115	*157	*199	*242
103	01 284	326	368	410	452	494	536	578	620	662
104	703	745	787	828	870	912	953	995	*036	*078
105	02 119	160	202	243	284	325	366	407	449	490
106	531	572	612	653	694	735	776	816	857	898
107	938	979	*019	*060	*100	*141	*181	*222	*262	*302
108	03 342	383	423	463	503	543	583	623	663	703
109	743	782	822	862	902	941	981	*021	*060	*100
110	04 139	179	218	258	297	336	376	415	454	493
111	532	571	610	650	689	727	766	805	844	883
112	922	961	999	*038	*077	*115	*154	*192	*231	*269
113	05 308	346	385	423	461	500	538	576	614	652
114	690	729	767	805	843	881	918	956	994	*032
115	06 070	108	145	183	221	258	296	333	371	408
116	446	483	521	558	595	633	670	707	744	781
117	819	856	893	930	967	*004	*041	*078	*115	*151
118	07 188	225	262	298	335	372	408	445	482	518
119	555	591	628	664	700	737	773	809	846	882
120	918	954	990	*027	*063	*099	*135	*171	*207	*243
121	08 279	314	350	386	422	458	493	529	565	600
122	636	672	707	743	778	814	849	884	920	955
123	991	*026	*061	*096	*132	*167	*202	*237	*272	*307
124	09 342	377	412	447	482	517	552	587	621	656
125	691	726	760	795	830	864	899	934	968	*003
126	10 037	072	106	140	175	209	243	278	312	346
127	380	415	449	483	517	551	585	619	653	687
128	721	755	789	823	857	890	924	958	992	*025
129	11 059	093	126	160	193	227	261	294	327	361
130	394	428	461	494	528	561	594	628	661	694
131	727	760	793	826	860	893	926	959	992	*024
132	12 057	090	123	156	189	222	254	287	320	352
133	385	418	450	483	516	548	581	613	646	678
134	710	743	775	808	840	872	905	937	969	*001
135	13 033	066	098	130	162	194	226	258	290	322
136	354	386	418	450	481	513	545	577	609	640
137	672	704	735	767	799	830	862	893	925	956
138	988	*019	*051	*082	*114	*145	*176	*208	*239	*270
139	14 301	333	364	395	426	457	489	520	551	582
140	613	644	675	706	737	768	799	829	860	891
141	922	953	983	*014	*045	*076	*106	*137	*168	*198
142	15 229	259	290	320	351	381	412	442	473	503
143	534	564	594	625	655	685	715	746	776	806
144	836	866	897	927	957	987	*017	*047	*077	*107
145	16 137	167	197	227	256	286	316	346	376	406
146	435	465	495	524	554	584	613	643	673	702
147	732	761	791	820	850	879	909	938	967	997
148	17 026	056	085	114	143	173	202	231	260	289
149	319	348	377	406	435	464	493	522	551	580
150	17 609	638	667	696	725	754	782	811	840	869
N	L O	1	2	3	4	5	6	7	8	9

Solution. For this problem we have $n = 5$, $i = .03$, and $S = \$300$. We use logarithms to compute $(1 + i)^n$:

$\log (1+i)^n = \log (1 + .03)^5 = 5 \log 1.03 = 5[0.01284] = 0.06420$.
$(1 + i)^n = $ antilog $0.06420 = 1.1593$.

Then

$$A = 300 \left[\frac{1.1593 - 1}{.03} \right] = 300 \frac{.1593}{.03} = \$1593.00. \quad Ans.$$

At the end of 5 years the man would collect $1593.00.

(b) What would be the amount of Example (a) if the interest were compounded semiannually?

Solution. We now have

$$j = .03 \quad \text{and} \quad m = 2$$

so that

$$(1 + j/m)^{mn} = (1.015)^{10} \quad \text{and} \quad (1 + j/m)^m = (1.015)^2$$

which are computed by logarithms:

$10 \log 1.015 = 10[0.00647] = 0.06470; \quad (1.015)^{10} = 1.1606$.
$2 \log 1.015 = 2[0.00647] = 0.01294; \quad (1.015)^2 = 1.0302$.

Then

$$A = 300 \left[\frac{.1606}{.0302} \right] = \$1595.36. \quad Ans.$$

Problem 2. *To find the present value of an annuity of* S *dollars per year, payable annually for* n *years.*

Let P be the present value, then

$$P = S \frac{1 - (1 + i)^{-n}}{i}.$$

If the interest is convertible m times a year at a yearly interest rate of j, then

$$P = S \frac{1 - (1 + j/m)^{-mn}}{(1 + j/m)^m - 1}.$$

Financial institutions have extensive tables to compute these quantities. However, some approximate values can be calculated by the use of five-place logarithms.

Example

A merchant places $1000.00 at the end of each year for 6 years into a sinking fund. Find the present value if money is worth 4% effective yearly interest.

Solution. We have

$$S = \$1000.00, \ i = .04, \ n = 6.$$

To compute $(1 + i)^{-n} = (1.04)^{-6}$ we use logarithms (it is necessary to have a complete 5-place table):

$$-6 \log 1.04 = -6[0.01703] = -0.10218 = 9.89782 - 10.$$
$$(1.04)^{-6} = \text{antilog}\,[9.89782 - 10] = 0.79035.$$

Then

$$P = 1000\,\frac{1 - .79035}{.04} = 1000\left(\frac{.20965}{.04}\right) = \$5241.25. \quad Ans.$$

The value of this annuity before the first payment is made is $5241.25.

Review Exercise VIII.

1. Find the simple interest on $1000:
   (a) for 5 years at $6\frac{1}{2}\%$;
   (b) for 48 days at $11\frac{1}{2}\%$ (ordinary);
   (c) for 9 months at 7%.

2. Find the compound amount on $1000 invested:
   (a) for 5 years at 6% compounded semiannually;
   (b) for 3 years at 4% compounded quarterly.

3. Find the discount and net proceeds on a 120-day note for $600 dated on June 1 and discounted on that date for $5\frac{1}{2}\%$.

4. What is the annual rate of interest on a loan of $250 which is paid through monthly payments of $17.97 for 15 months?

5. Find the present worth of $1000 due in 5 years with interest at 5% compounded semiannually.

6. Find the amount of savings if $500 is put into a bank at the end of every year for 5 years and the bank pays 4% compounded semiannually.

7. How long will it take $100 to double itself if invested at 6% compounded semiannually?

Chapter X

MATHEMATICS OF BUSINESS

Introduction

In the United States business is conducted on the basis of a fair profit. In industry, raw materials are collected and fabricated into products which are sold, or certain parts may be purchased and assembled into new products and then sold. In commerce, merchants and storekeepers buy goods from manufacturers and sell them to consumers. All along the line certain labors are applied and each is paid for his efforts through the profit system. Business, then, is concerned with buying and selling, with application of labor and overhead, and finally with a fair profit.

Buying

A merchant starts his business by first obtaining goods which possess the quality he desires to distribute to his customers. He usually purchases these goods from various manufacturers or an intermediary called a wholesaler. The purchases are often made from a catalog or price sheets which give the specifications and the prices. This price is called a *wholesale price*. It is what the merchant pays for his goods, but it is not necessarily the final price since the manufacturer or wholesaler may give *trade discounts*. Such discounts are usually for cash paid within a certain date or for volume buying. We have already discussed (see Chapter III) how to compute these discounts.

In addition to the wholesale price that a merchant pays, there are usually added transportation charges in order to get the goods moved from warehouses to the store. This cost is a fixed charge dependent upon the mode of transportation and weight.

Example

A merchant bought 12 cameras listed in the catalog at $28.50 with discounts of 20% and 10% and a transportation charge of $0.50 per camera. How much did the cameras cost him?

Solution.

List price:	(12)($28.50)	$342.00
Less 20%:	($342.00)(.20)	68.40
		$273.60
Less 10%:	($273.60)(.10)	27.36
		$246.24
Plus Transportation:	($0.50)(12)	6.00
Total Cost:		$252.24 *Ans.*

Modern business requires that accurate records be kept of all transactions. Thus in the buying from whosesalers or manufacturers, *invoices* are submitted with each order. These invoices are usually in the form of an itemized bill and also state the terms of the sale. The terms are frequently in abbreviated form, and here are some of the abbreviations used:

Terms: 3/10, n/30 means "a discount of 3% allowed if paid within 10 days; if not, full amount is due 30 days after date of bill."

Terms: 3/10, 1/30, n/60 means "3% off if paid within 10 days, 1% off if paid within 30 days, net amount in 60 days."

Terms: 3/10 EOM means "3% off if paid within 10 days after end of month in which bill is received."

Example

FOOD DISTRIBUTORS, INC.

133 Fifth Ave., Devon, Pa.

Sold to: Date _____

Hams Food Market
3333 Locust St.   Terms: 3/10, n/30
Farmington, Pa.

---

3 cases	No. 3 Peas	at 3.40	10.20
4 cases	No. 2 Peaches	at 3.65	14.60
5 cases	No. 10 Corn	at 2.90	14.50
10 cases	No. 4 Beets	at 3.10	31.00
	Total		70.30
	Freight		2.50
	Total		72.80

What is the amount of the check to be sent to Food Distributors, Inc. if bill is paid in 5 days?

Solution. The terms are 3% discount within 10 days. The discount is figured on cost of the goods (not including freight charges), thus:

Discount = $70.30 (.03) = $2.11.

Amount due = $72.80 − $2.11 = $70.69. *Ans.*

Overhead and Profit

In the conduct of his business, a merchant has other expenses besides the cost he must pay for his goods. These expenses include rent, utilities, wages to employees, supplies, operating equipment such as trucks for delivery services, advertising, etc. Such expenses are usually grouped into an item called *overhead*. Before we consider the mathematics involved let us first define a few terms.

Cost (net)   is usually the amount of money a merchant pays for an article.

Overhead   is the expenses necessary to operate the business.

Gross Cost   is the sum of the net cost and the overhead.

Selling Price   is the amount of money for which the article is sold.

Profit   is the net earnings or the difference between the selling price and the gross cost.

Margin   is the sum of the overhead and the profit or the difference between the net cost and the selling price.

Mark-Up   is the amount that is added to the net cost to provide for the overhead and the profit. It is the same as the margin.

Market Price   is the price that appears on the price tag for the article; it is the same as the selling price if no discount is allowed.

We can now write the mathematical relations among the terms.

Selling Price	= cost + margin
	= cost + overhead + profit
	= market price − discount
Margin	= overhead + profit
	= selling price − cost
Profit	= selling price − gross cost
	= margin − overhead
Gross Cost	= net cost + overhead

In discussing a business the quantities for profit and margin are often expressed in terms of per cent, which is obtained by considering the ratio of the quantity to the selling price; for per cent of mark-up we form the ratio of the mark-up to the net cost.

Examples

(a) A dealer bought a television set for $122.00 and paid shipping charges of $10.00. He sold the set for $287.50. What was the margin, the mark-up, the per cent of margin, and the per cent of mark-up?

Solution.

Invoice Cost: $122.00	Selling Price: $287.50
Freight: 10.00	Net Cost: 132.00
Net Cost: $132.00	Margin: $155.50

The margin and mark-up are the same in amount.

Per cent of Margin: $\frac{155.50}{287.50} = .54 = 54\%$.

Per cent of Mark-Up: $\frac{155.50}{132} = 1.18 = 118\%$.

(b) The books of a gift shop showed that during the month of July it had purchased $9,875.50 worth of goods and had sold them for $22,980.00. The books also showed expenses of

Rent	$425.00	Supplies	$ 8.75
Wages	980.00	Advertising	36.00
Electricity	54.00	Repairs	42.50
Telephone	15.60	Miscellaneous	7.80

Itemize the following: Net Cost, Overhead, Gross Cost, Total Selling Price, Margin, Profit, and the per cent of each.
Solution.

Item	Formula	Amount	%
Selling Price	Given	$22,980.00	100
Net Cost	Given	$ 9,875.50	43.0
Overhead	Total of Expenses	$ 1,569.65	6.8
Gross Cost	Net Cost + Overhead	$11,445.15	49.8
Margin	Selling Price — Net Cost	$13,104.50	57.0
Profit	Selling Price — Gross Cost	$11,534.85	50.2

To obtain the per cent, divide the amount of each item by the selling price, $22,980.00 and multiply by 100. The selling price is, of course, 100% and since it is made up of the cost of the goods plus the overhead plus the profit, we must have the sum of the last three total 100%.

Selling Price = Cost + Overhead + Profit:

$$100\% = 43\% + 6.8\% + 50.2\%.$$

The selling price is often established by first considering the desired per cent of the net profit. In other words, if the cost of an item and the overhead in handling it are given, then by first picking the per cent of profit desired, the selling price can be established. Let x be the selling price and let y be the desired per cent of profit, then

$$1.00x = \text{Amount of Cost} + \text{Amount of Overhead} + (.0y)x$$

or

$$(1 - .0y)x = \text{Gross Cost}.$$

Example

An air conditioning contractor pays $450 for a unit and it costs him $250 to sell and install it. What should be the selling price if he desires to make a 35% profit?

Solution.

$$\text{Gross Cost} = \$450 + \$250 = \$700.$$
$$(1 - .35)x = 700.$$
$$x = \frac{700}{.65} = \$1076.92. \quad Ans.$$

Frequently merchants establish their per cent of profit so that in case a certain item fails to sell, they can run a special sale with a discount and still make a profit.

Example

A merchant established a market price of $10.98 on an item which cost him $4.25 and had a rated overhead of $0.75. He then ran a sale on the item marked "$\frac{1}{3}$ off." What were the actual selling price, his original per cent of profit, and his actual per cent of profit?

Solution.

Sale Price = $\frac{2}{3}$($10.98) = $7.32.

Gross Cost = $4.25 + $0.75 = $5.00.

$$\text{Original \% of Profit} = \frac{10.98 - 5.00}{10.98} = \frac{5.98}{10.98} = 54\%.$$

$$\text{Actual \% of Profit} = \frac{7.32 - 5.00}{7.32} = \frac{2.32}{7.32} = 32\%.$$

The overhead for a store can be "pro-rated" down to a set of articles or even down to each item. There are a number of ways in which this can be done. For a small business it can be done as a percentage of sales.

Example

For the month of July a hardware store has a total overhead of $2,913.12 and a total sales of $24,276.00. Kitchenware sales amounted to $1,280.00. How much of the overhead should be charged to kitchenware?

Solution.

$$\text{Per cent of Overhead} = \frac{2913.12}{24276.00} = .12 = 12\%.$$

Kitchenware: $(1280.00)(.12) = \$153.60$. *Ans.*

Today, even a small business requires a considerable amount of bookkeeping, and a considerable amount of mathematics is involved. Such mathematics forms the subject of accounting and should be studied as such.*

Review Exercise IX.

1. A merchant bought the following items from a wholesaler:

Quantity	Unit Price	Discount
12	$3.50	20%, 10%
144	$0.50	30%
50	$1.80	10%, 5%
25	$0.60	10%, 3%

How much did he pay?

2. A merchant has the following expenses:

Rent — $500.00; Wages — $1200.00; Utilities — $112.00; Supplies — $37.50; Advertising — $88.00; Repairs — $110.00; Deliveries — $92.00; Miscellaneous — $8.25. He sold goods which cost him $11,800.00 for $25,750.50. Find the following: Net cost, overhead, gross cost, total selling price, margin, profit, and the per cent of each.

* See R. D. M. Bauer and P. H. Darby, *Elementary Accounting* (New York: Barnes & Noble, 1952).

3. Establish the selling price for an item costing (gross cost) $1.25 to obtain a 40% profit. If the item was then sold at a 10% discount, what was the actual per cent of profit?

4. A Fur Salon ran a summer sale, "40% off." If the mark-up on all items was 110%, what was the actual per cent of profit? (Hint: Let the cost be $1.00.)

5. Under the conditions of problem 4, what was the "sale" price of a fur coat which cost the merchant $600.00?

Chapter XI

MATHEMATICS OF LIFE

Introduction

The everyday affairs of living involve the use of mathematics more and more. The once simple process of determining your salary now involves a complete tabulation of deductions. Buying a home, the family car, insurance, and other installment buying practically makes each family a business concern. We shall now turn our attention to a discussion of some of the more fundamental aspects of the mathematics of life.

Income

Most people's income stems from wages or salary in return for their labors; however, any money received that is not a gift or inheritance is considered income. A wage earner may be paid so much per hour, per day, per week, per month, or per year; he may be paid a commission or so much per piece. He may receive a pay check every week, every two weeks, twice a month, or every month.

The normal working week has changed over the years and at this writing it is usually 8 hours per day 5 days a week for a 40 hour week. If a worker works more than 40 hours a week, he is usually paid for the *overtime* at a premium rate.

Examples

(a) A man is paid $1.70 an hour and a premium rate of $1\frac{1}{2}$ times for any work over 40 hours per week. How much did he earn if he worked 48 hours?

Solution.

$$40 \text{ hours at } \$1.70/\text{hr.} = \$68.00$$
$$8 \text{ hours at } \$2.55/\text{hr.} = \$20.40$$
$$\text{Total} = \$88.40. \quad Ans.$$

(b) A carpenter is paid $2.80 per hour, an overtime rate of $1\frac{1}{2}$ for any overtime during the Monday-Saturday time, and double pay for Sunday work. How much did a carpenter earn if he worked the following hours:

Days Hrs.	Mon. 10	Tues. 10	Wed. 10	Thurs. 10	Fri. 8	Sat. 4	Sun. 4

Solution.

Regular hours: 40 at 2.80 = $112.00
Time & Half hours: 12 at 4.20 =   50.40
Double Time hours: 4 at 5.60 =   22.40
            Total   = $184.80.

(c) A professional man is hired at an annual salary of $7,280. How much is his pay check if he is paid twice a month? Every two weeks?

Solution.

Twice a month would call for 24 pay checks a year so that each pay check would be

$$\tfrac{7280}{24} = \$303.33. \quad Ans.$$

Every two weeks would call for 26 pay checks so that each pay check would be

$$\tfrac{7280}{26} = \$280.00. \quad Ans.$$

A wage earner can obtain his income in a number of ways: the salary as described above, commissions, fees, etc. We shall list the most common ones:

Straight Salary — so much per hour
Overtime Salary — premium rate
Piecework — so much for each piece made
Straight Commission — so much per item
Graduated Commission — increasing rate for more sales
Salary and Commission — fixed base plus commission
Bonus — extra pay for success
Fees — agents' and brokers' charges
Professional Fees — doctors, lawyers, dentists, etc.
Rents and Royalties — house rents, inventions, authors

Examples

(a) A sales girl in a department store receives a salary of $30 per week plus 5% on all sales over $500. How much did she earn in a week in which her sales totaled $850?

Solution.

> Sales over minimum: $850 - $500 = $350.
> Commission: $(350)(.05) = 17.50
> Straight Salary: $ = 30.00
>   Total = $47.50. *Ans.*

(b) A dentist charges $3.00 for each X ray, $6.00 for cleaning teeth, $4.00 a filling, and various fees for bridge work. What was his gross income (not including bridge work) if he had 200 X rays, 200 cleaning jobs, and 300 fillings during January?

Solution.

> X rays: $(200)($3.00) = $ 600$
> Cleanings: $(200)($6.00) = 1200
> Fillings: $(300)($4.00) = 1200
>   Total = $3000. *Ans.*

(c) A real estate agency receives a commission of 5% of the sale of each house. The salesman gets $2\frac{1}{2}\%$ and the rest goes to the agency. If Mr. Jones sold $300,000 worth of real estate, what was his income?

Solution.

> $($300,000)(.025) = $7500.00. *Ans.*$

Income is also received from investments such as interest from a bank account, dividends from stocks and bonds, and the sale and purchase of stocks and bonds.

Example

Mr. Jones works for $2.25 per hour, and last year he worked 2080 hours. He has a bank savings account of $3500 which pays 3% compounded semiannually, and owns $1000 of utility bonds at 4% and 200 shares of stock which pays $2.00 per share annual dividends. What was his income for last year?

Solution.

> Annual Salary: $(2080)(2.25)$ $ = 4680.00
> Interest (Table V): $(3500)(1.0302) - 3500 =$ 105.70
> Bonds: $(1000)(.04)$ $ =$ 40.00
> Stocks: $(200)(2.00)$ $ =$ 400.00
>   Total = $5225.70. *Ans.*

Take-Home Pay

The average man in the United States who works for wages will have an income as discussed in the preceding section. However, this will not necessarily be his "take-home" pay since certain deductions are made by the company. We shall discuss some of the most common ones.

Federal Income Tax. The present federal tax law requires that the employer deduct the income tax from each pay check. The amount to be deducted depends upon the salary bracket and the number of dependents that an employee has. Each employee files with his employer a statement of the number of dependents; the employer then establishes the percentage of deduction depending upon the individual income. The resulting amount is then deducted from the pay check, and the employer sends that amount to the Internal Revenue Service. Adjustments are made when the individual files his income tax return.

Social Security (F.I.C.A.). In August, 1935, the United States passed an act to provide an income for the aged. This act has undergone many revisions, and the amounts are continually changing as the economic conditions of the country change. The purpose of the act is to provide some income for retired people and their dependents. The money is collected by an equal contribution from the employee and the employer, which is then put into a special trust fund from which all benefits are paid. The contributions are based on an established percentage up to a maximum annual amount. (For complete details on current contributions and qualifications for benefits, the reader should contact the local social security office and obtain its current literature.)

State or City Tax. Some states and some cities have local income taxes which are deducted by the employer and sent to the proper authorities. These are also based on a certain percentage of the income.

Insurance. Most large companies have insurance plans to cover hospitalization, sickness, accident, and life. These plans are usually based on a contribution by employee and a contribution by the employer. The benefits depend upon the particular plan that the company is associated with. The employee's contribution is usually a specific amount which depends upon the amount of coverage that he selects.

Union Dues. If an employee falls into a category where he belongs to a trade union, the contract between the union and the company usually provides for the deduction of the union dues from the employee's salary. These dues are a fixed amount depending upon the employee's job classification.

Contributions. Most employers will provide a payroll deduction service for contributions to charity organizations. The employee specifies the amount and the organization to which the contribution should be sent.

Bonds. Employees can purchase government bonds on a payroll deduction plan. Again the employee specifies the amount, which is usually some fraction of the purchase price of the bond.

Retirement Programs. Some of the larger corporations have a retirement program over and above the social security program. Such programs usually supplement the social security plan and are based on joint contributions by the employee and the employer. The amount of the contributions are figured on a percentage basis of the employee's salary and are deducted from each pay check.

Many of the above programs are voluntary, and the employee must elect to participate. Since the percentages used for deduction purposes vary from company to company, vary according to the particular plan in use and tax laws, and vary according to the salary category of the employee, definite amounts cannot be stated. Details of all plans are usually obtained from the company's personnel office. Some typical examples will be given, but it must be remembered that these are only representative numbers.

Examples

(a) Mr. Jones works as a machinist at a rate of $3.25 per hour and gets paid every two weeks. He participates in various plans, and the following deductions have been established: Federal Income Tax — 15%, State Tax — 1%, F.I.C.A. — $4\frac{1}{2}\%$, and for each pay-check period Union Dues — $5.00, Insurance — $4.50, Contributions — $2.00, Bonds — $6.25. What is his take-home pay?

Solution:

Earnings:	(80)($3.25)		$260.00
Deductions:			
Income Tax:	(260)(.15)	39.00	
State Tax:	(260)(.01)	2.60	
F.I.C.A.:	(260)(.045)	11.70	
Union Dues:		5.00	
Insurance:		4.50	
Contributions:		2.00	
Bonds:		6.25	
	Total Deductions:		$ 71.05
	Take-Home Pay:		$188.95 *Ans.*

(b) Mr. Smith has an annual salary of $12,000.00 and gets paid every month. His deductions are as follows: Income Tax — 18%, State Tax — 2%, F.I.C.A. — $4\frac{1}{2}$%, Insurance — $12.00, Contributions — $5.00, Retirement — $80.00, Bonds — $75.00. What is the amount of his check?

Solution.

Earnings:			$1000.00
Deductions:			
Income Tax:	(1000)(.18)	$180.00	
State Tax:	(1000)(.02)	20.00	
F.I.C.A.:	(1000)(.045)	45.00	
Insurance:		12.00	
Contributions:		5.00	
Retirement:		80.00	
Bonds:		75.00	
Total Deductions:			$ 417.00
Amount of Check:			$ 583.00 *Ans.*

Banking

Now that we have an income and a regular take-home pay, we should have a place to keep our money, and our society provides us with an excellent place, namely, a bank. Many services can be obtained from a bank, and most responsible people make extensive use of the banks. The two most common services are the savings account and checking account, each of which is started by simply making a deposit. When an account is opened, a *passbook* is issued in the name (or names) of the person who will control that account; that is, the person (or persons) who will make deposits and withdrawals from that account. To make a deposit, the money, passbook, and a deposit slip are turned over to a *teller* of the bank. The teller makes the entry in the passbook, which is a record for the owner of the account, and files the deposit slip, which is a record for the bank. A typical deposit slip is shown in Figure 49, p. 174.

Most business transactions are made by checks, and many people use checks in paying their bills. Not only are checks convenient but they also become recognized receipts and aid in individual bookkeeping. A typical check is shown in Figure 50, p. 175.

A check is made out to a *payee*, the person (or firm) who is to receive the money. The payee can convert the check into cash by signing his name on the back of the check; this is called *endorsing* the check.

FOR DEPOSIT TO CHECKING ACCOUNT OF

Ralph Roe

ADDRESS *110 West Third St.*
(IF THIS IS A NEW ADDRESS PLEASE NOTIFY TELLER)

WITH THE

FIRST NATIONAL BANK
AND TRUST COMPANY

UPON THE TERMS AND CONDITIONS OF THE
AGREEMENT PRINTED ON THE BACK HEREOF

DOBBSVILLE, OHIO *Jan 10* 19*64*

	DOLLARS	CENTS
CURRENCY		
COIN		
CHECKS (List bank number and amount of each check separately.)		
BANK NO. 20-1/712	139	50
" " 14-1/810	476	80
" "		
" "		
" "		
" "		
" "		
ACKNOWLEDGE RECEIPT OF CASH RETURNED BY SIGNING BELOW — TOTAL	616	30
LESS CASH RETURNED	100	00
SIGNATURE — NET DEPOSIT	516	30

F-204 ARE ALL CHECKS ENDORSED?

Fig. 49

The payee can transfer the check to another party by endorsing it in full and placing the name of the second party on the back of the check.

FIRST NATIONAL BANK
AND TRUST COMPANY
NO. *315*

DOBBSVILLE, OHIO *June 16* 19*61* 20-1/712

PAY TO THE
ORDER OF *Acme Investment Co* $ *150.00*

One Hundred and Fifty and 40/100 DOLLARS

Ralph Roe

Fig. 50

The payee can also restrict the endorsement by specifying the purpose
on the back of the check.　Examples are shown in Figure 51.

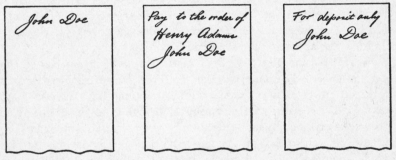

Fig. 51

Checks are drawn against money which has already been placed in
the bank.　At the present time there are two prominent types of
checking accounts: one requires a certain minimum balance and
charges according to the smallest balance and the number of checks

No.	Date	Payee	Amt.	Deposit	✓	Balance	
314	6/1	Local Mortgage Co.	134.17			1359	12
315	6/16	Acme Inv. Co.	150.00	200.00		1409	12
316	6/19	Joe's Garage	37.15			1371	97
317	6/21	W. C. Eyre Co.	138.85		✓	1233	12

Fig. 52

written during a month; the other furnishes a book of checks at so much per check. An accurate account of the balance in a checking account should always be maintained. This can best be accomplished by making an entry in the check stubs (or record sheet) for each check. These records show the balance at each date and will prevent over-drawing or falling below a required minimum, thus saving you money. A typical record sheet is shown in Figure 52, p. 175. The reader should check the numbers for each entry.

Banks offer many other services such as Christmas Funds, Loans, Annuities, Estates, etc.

Taxes

With a good job that provides us with an income and a bank in which to keep our money, let us now take a look at some of our expenses. We shall first consider taxes, of which today there are many; however, they may be put into the following categories: (1) income taxes; (2) property taxes; (3) excise and sales taxes; (4) customs; and (5) special taxes. We shall consider each in turn.

Federal Income Tax. We have already met the income tax in the calculation of take-home pay; however, we are not required to pay a tax on our gross income as certain exemptions and deductions are permitted. The rates and exemptions change from time to time as they are prescribed by law. These laws become very complex but are usually written so that an average individual can figure his own tax. Each individual who has an income over a specified minimum is required to file with the Internal Revenue Service an "Income Tax Return." This is a statement of his gross income, deductions, net income, calculated tax, and withholdings, and should be accompanied by a check for the balance due, if any.

We shall present an example based on the tax form for a recent year. Again it should be remembered that this is only an example, and we are not covering all the details. The tax return form itself has excellent instructions; for more complicated cases, it is advisable to consult an attorney. We shall consider only the first two pages of a return and give the example under the following conditions:

(a) Married couple with two dependent children and filing a joint return;
(b) All income from salaries and wages;
(c) Itemized deductions.

I. *Identification.*

Name, address, social security number, and occupation of both husband and wife.

II. *Exemptions.*

Blocks are provided to check exemptions for yourself, wife, and other dependents with their names and addresses. Then on line 4 enter the total number of exemptions claimed which in our example will be 4.

III. *Income.*

Enter all wages, salaries, etc.

Employer's Name	Where (City & State)	(a) Wages	(b) Income Tax Withheld
Acme Corporation	Basetown, N.Y.	$ 9800.00	$1372.00
Jeanies Salon	Basetown, N.Y.	2400.00	312.00
	Totals	$12200.00	$1684.00

Lines 6–10 deal with sick pay and profit or loss from other sources. Line 11 is the adjusted gross income tax, which in our example is $12,200.00.

IV. *Deductions.*

We shall itemize the deductions in order to show typical cases. These are listed on page 2 of the return.

(a) Contributions

Second Church	$250.00	
United Charity Fund	125.00	
Boy Scouts	30.00	
Total		$ 405.00

(b) Interest

Home Bldg. & Loan Assn.	$632.55	
National Bank (Loan)	35.00	
Total		$ 667.55

(c) Taxes

Real Estate Taxes	$445.25	
State Income Tax	112.00	
Personal Property Tax	82.00	
State Gasoline Tax	87.50	
Auto License	12.25	
Total		$ 739.00

(d) Medical & Dental Expenses

Less than 3% of line 11   000.00

(e) Other Deductions

AAA Society Dues $ 9.00

Total $ 9.00

Total Deductions $1820.55

V. *Tax Computation*

1. Adjusted Gross Income $12200.00
2. Deductions   1820.55
3. Line 1 less line 2 $10379.45
4. Exemptions ($600 × 4)   2400.00
5. Taxable Income   7979.45
6. Tax   1675.48

The tax is computed from a tax schedule given with the tax return form, and we have used Schedule II for married taxpayers filing joint returns. The first part reads:

If your taxable income is: Your tax is:

Not over $4,000.......   20% of your taxable income

Over   But not over   *of excess over —*

$4,000– $8,000........ $800, plus 22%–$4,000

$8,000–$12,000........ $1680, plus 26%–$8,000

etc.

Thus for our case we have

$$\$800 + \$(3979.45)(.22) = \$1675.48.$$

VI. *Tax Due.*

We return to page 1 of the return where the tax is entered on line 12. Lines 13–15 deal with credits from dividends, retirement, and self-employment tax. Thus we have

16.   (for us same amount as line 12) $1675.48
17.(a) Tax withheld $1684.00
   (b) Payments on Estimated Tax   0.00 $1684.00
18.   Balance due
19.   Overpayment $ 8.52

Since we had an overpayment, this amount is refunded by the government or credited to next year's tax if we prefer.

State Income Tax. Many of the states now have income taxes, and the forms for returns are usually simpler than the federal income tax forms. Certain exemptions and deductions are allowed, and the rates are quoted in terms of per cent of the taxable income.

Property Taxes. Everyone who owns property pays property taxes to the state and local governments. These taxes are used to pay the government operating expenses for official police and fire protection, schools, government buildings, streets, sanitation, etc. We shall now discuss the manner in which these taxes are usually calculated.

First all property is assessed by a special agency known as the *tax assessors;* this means that the property is given a value or a price is established by the assessors. The amount is usually a fraction of the market value and may change from year to year depending upon the property (such as automobiles) or may remain fixed for a certain period of time. Then each year the state, the county, the township, and the city establish their budgets and estimate the amount of money they need for that year. Now knowing the amount of money needed and the total assessed value of the property within their area, each group establishes a *tax rate* by simply dividing the total property value by the amount needed.

Let us consider a city budget and for convenience let us group many items into some bigger units.

Annual Budget Estimate, 19—

Expenditures

Department of Education	$ 3,500,000
Department of Safety	1,675,000
Public Works	2,300,000
Department of Health	1,200,000
Administration	975,000
Buildings and Grounds	1,600,000
Payments on Bond Issues	2,150,000
Taxes	500,000
Total	$13,900,000

Revenues

Direct Payments for Services	1,400,000
Amount Needed	$12,500,000

Let us now suppose that the total assessed value of all property in the city is $200,000,000. The tax rate for this year is then

$$\frac{12,500,000}{200,000,000} = .0625.$$

This tax rate can be expressed as so much per dollar of valuation, per \$100 of valuation, or per \$1000 of valuation.   Thus in the above example we would have

> 6.25¢ on the dollar;
> \$6.25 on \$100 valuation;
> \$62.50 on \$1000 valuation.

Now suppose that we own a house and lot in this city which the tax assessor has assessed at \$8200.   We would have to pay

$$\tfrac{8200}{100}(6.25) = \$512.50$$

property tax.

In many localities there are two kinds of property taxes:

(a) Real estate — land and permanent improvements such as houses, garages, etc.

(b) Personal property — furniture, automobiles, appliances, etc.

Both of these taxes are calculated the same way but are usually paid separately.

Excise and Sales Taxes.   These taxes are levied on goods which are sold, and there are both federal and state taxes involved.   The federal taxes may be collected at the manufacturer or at the retail sale.   Most state taxes are collected at the retail sale by the store.   For example, there is a federal and state tax on gasoline which is quoted at so much per gallon.   The amount is added into the sales price of gasoline and the consumer simply pays the gasoline retailer, who in turn apportions the money.

Example

The federal tax on jewelry is 10% and there is a state sales tax of 3%.   How much is paid for a watch which has a retail price of \$67.50?

Retail Price	=	\$67.50
Federal Tax $(67.50)(.10)$ =		6.75
State Tax $(67.50)(.03)$	=	2.03 *
Total Price	=	\$76.28 *Ans.*

Customs Taxes.   Commodities which are imported from foreign countries are usually subject to taxes which are called *customs taxes* or *duties*.   If no tax is levied on an imported article, it is said to be *duty-free*.   The tax is specified in either of two ways:

---

* Taxes are usually stated so that the collections are rounded up, i.e., (67.50)(.03) = 2.025 is rounded off to 2.03.

(a) *ad valorem duty* — a per cent of the net value;

(b) *specific duty* — so much per quantity.

Examples

(a) What is the duty on 12 cameras valued at $29.80 each and taxed at 35% of value?

Solution.

Value: ($29.80)(12) = $357.60.
Tax: ($357.60)(.35) = $125.16. *Ans.*

(b) What is the duty on 100 tons of sugar taxed at $1\frac{1}{2}$ cents per pound?

Solution.

Quantity: (100)(2000)   = 200,000 lbs.
Tax: (200,000)(.015) = $3000.00. *Ans.*

In some cases the government levies both types of taxes on imported goods and the amount paid is the sum of the two.

Special Taxes. There are many special taxes which are levied by the federal, state, and local governments. Some of these are in the form of licenses and fees; excise taxes on liquor, tobacco, etc.; tolls on bridges and roads; recording of legal documents; estate and inheritance taxes; and many others. Many of these are in the form of fees for services rendered by governmental agencies.

Home Ownership

Lodging is another of our expenses. We can either rent or own our home. If we rent, the rate is usually established at so much per month. If we own our home, there are a number of expenses to be accounted for. It is difficult to say which is the cheaper as it depends upon the individual taste. In this section we shall discuss the mathematics connected with owning a home.

Let us begin by buying a house. We could either buy a lot and contract with a builder to build a home to our specifications, or we could buy a house that has already been constructed, in which case the lot is a part of the sale. We decide to buy a house which a real estate agency has listed for $22,500.00. We make a proposition to the owner through his agent to purchase the house for $21,000.00 contingent on our ability to obtain a mortgage of $16,000.00 on the property and deposit $1000.00 as earnest money. The owner accepts the offer. We now apply to a loan agency or bank for the mortgage, and while the paper work is in progress we obtain an abstract of the title to the property from the real estate agent. It is desirable to retain the services of a lawyer to review the abstract in order to be certain that the title is clear and that there are no liens against the property. We are successful in our mortgage application and obtain a mortgage from a bank at $4\frac{1}{2}\%$ for 18 years. Our lawyer informs us that the title is clear and advises us to purchase the property. A closing date for the purchase is set. The transaction involves:

(1) To be paid to the present owner:

Earnest Money (already paid by us)	$ 1,000.00
Mortgage (paid by our bank)	16,000.00
Balance of Price (paid by us)	4,000.00
Total	$21,000.00

(2) To be paid to the Real Estate Agent:

Commission of 5% (paid by present owner):

($21,000)(.05) = $1050.00

(3) To be paid to our lawyer:

Services (paid by us): $65.00

(4) To be paid to our bank:

Closing costs on mortgage (paid by us): $175.00

(5) Delivery of deed to us by the present owner, thereby giving us the title to the property.

The cost to us at this time is

Earnest Money	$1,000.00
Balance of Down Payment	4,000.00
Legal Fees	65.00
Mortgage Fees	175.00
Total	$5,240.00

The above figures are representative, depending upon the fees of the lawyer and the bank, and should not be considered fixed amounts.

We now own a house and make plans to move into our new home. At this time most people buy some new furniture, rugs or carpets, curtains or drapes, and do some interior decorating. The amount spent depends upon what is needed. Let us assume that we spend $1000.00. The total initial cost of moving into our home is then $5,240.00 + $1000.00 = $6,240.00.

We shall now consider the annual expenses involved in owning a home.

A. *Mortgage.* A mortgage is a conditional transfer of property as security for a loan. In the purchase of our house we borrowed $16,000.00 from a bank and promised to pay this amount plus interest at the rate of $4\frac{1}{2}\%$ in 18 years. If we fail to meet our obligations, the bank has a right to *foreclose* and assume ownership of the property. Although there are many ways in which the loan may be paid, the most widespread practice today is to make regular payments, of which a certain amount is for the interest and a certain amount is for the reduction of the principal. This is called a process of *amortization*. Since the principal is reduced with each payment, the interest for the next payment is calculated on the balance. If a fixed amount is paid at each payment, the amount for the interest will diminish and the amount for reduction of the principal will increase. Payments are usually set for each month, and the amount of the payment can be established from amortization tables. The loan company provides a pass book in which entries are made at each payment, showing the date due, date paid, amount applied to the interest, amount applied to the principal, and balance due. For the mortgage obtained in the example we have been discussing, the monthly payments are $101.23. The first five entries in our passbook will read:

Date	Amt.	Interest	Principal	Balance
—	101.23	60.00	41.23	15,958.77
—	101.23	59.85	41.38	15,917.39
—	101.23	59.69	41.54	15,875.85
—	101.23	59.53	41.70	15,834.15
—	101.23	59.38	41.85	15,792.30

After 26 payments the entries will read:

—	101.23	55.96	45.27	14,876.25
—	101.23	55.79	45.44	14,830.81

A few years "down the road":

—	101.23	21.85	79.38	5,748.53
—	101.23	21.56	79.67	5,668.86

B. *Insurance.* Most home owners protect their investment against fire, theft, wind storms, etc. by carrying insurance. This is usually divided into three main categories: insurance on the house and grounds, insurance on the furnishings, and property liability insurance.   Most companies will offer all three as a package deal.   The premiums are based on the value placed on the property and are usually quoted at so much per $1000.00.   In a package policy the value is placed on the buildings and the furnishings are then insured up to a percentage of the main value.   Let us insure our new home for $18,000.00 and obtain a policy which also insures our furnishings up to 40% of the value placed on the house.   Thus our furniture is insured for a maximum of

$$(\$18,000.00)(.40) = \$7,200.00.$$

If the premium for this policy is quoted at $3.67 per thousand, we would pay

$$\left(\frac{18,000}{1000}\right)(\$3.67) = \$66.06$$

as our annual premium.

C. *Taxes.*   We have already discussed property taxes in Section 77. Let us assume that our house is assessed at $6800.00 and that the tax rate has been set at $6.12 per $100 evaluation.   Our real estate tax is

$$(\tfrac{6800}{100})(\$6.12) = \$416.16$$

for this year.

If our household furnishings are assessed at $400.00 and the tax rate is the same, then our personal property tax is

$$\tfrac{400}{100}(\$6.12) = \$24.48.$$

D. *Utilities.*   Another monthly expense stems from our use of public utilities such as electricity, water, gas, and telephone.   In some communities there are also charges for sewers and for trash collection. Let us say that our average monthly costs are as follows.

Electricity	$12.75
Gas	7.50
Telephone	5.60
Water	4.25
Total	$30.10

E. *Repairs and Improvements.*   Everyone who owns a home has a certain pride of ownership and a desire to keep the property in good shape.   This necessitates repairs and improvements as material wears out.   Some savings can be effected by a "do-it-yourself" approach. On a "do-it-yourself" project considerable time can be saved and trouble prevented by planning the job.   Let us consider a couple of examples for our recently acquired home.

(a) We desire to paint one of the bedrooms. The bedroom measures 12′ 3″ by 14′ 6″ and the ceiling is 8′ from the floor. The total wall surface is

$$2(12.25)(8) = 196.0$$
$$2(14.50)(8) = 232.0$$
$$\text{Total} = \overline{428.0} \text{ sq. ft.}$$

In calculating this surface we made no allowances for the openings due to doors and windows, so the area to be painted will be less than the above figure. The manufacturer of the paint we wish to buy states on the can that a gallon will cover 425 sq. ft. Thus we purchase one gallon at a cost of $4.75 per gallon. We will also need a brush or two. We decide to purchase one large brush for the flat wall area and one small brush for close work around windows and corners. The cost of the brushes is $3.75 and $2.50. By proper care we will be able to use these brushes on other painting jobs. Thus the cost of painting the bedroom is

$$\$4.75 + \$3.75 + \$2.50 = \$11.00.$$

We have not charged for our own time in doing the job; this represents our savings. If, however, we take time off from our regular job and thus lose some wages, then we must add that to the cost of painting the bedroom.

(b) The second example will deal with the laying of a new sidewalk along the side of our garage and along the back to the back door. We make a sketch of the layout and measure the distances. The result is shown in Figure 53. The total area is

$$(28')(3') + (40')(3') = 204 \text{ sq. ft.}$$

Sidewalks are usually 4″ thick so that the total volume is

$$(204)(\tfrac{4}{12}) = 68 \text{ cu. ft.}$$

"Readi-Mix" concrete is usually sold by the cubic yard and can be bought by the $\frac{1}{2}$ "yard." Since there are 27 cubic feet in one cubic yard we need

$$\tfrac{68}{27} = 2.52 \text{ cubic yards.}$$

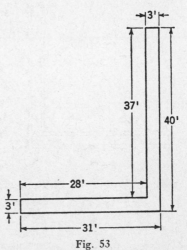

Fig. 53

Thus, $2\frac{1}{2}$ cubic yards will be slightly short, but by making the sidewalk a little less than 4″ thick in spots we can make $2\frac{1}{2}$ "yards" do. Let us say that the concrete costs $14.75 a "yard." The cost of the concrete will be

$$(\$14.75)(2\tfrac{1}{2}) = \$36.88.$$

To make the forms we need some 1×4 lumber and some 2×2 stakes. We can purchase the 1×4 in 8′, 12′, and 16′ lengths and the 2×2 in 8′ lengths. The perimeter of our form is

$$40' + 3' + 37' + 28' + 3' + 31' = 142'.$$

To fit this perimeter we could use

$$16 + 16 + 8 \text{ to get the 40,}$$
$$16 + 16 + 8 \text{ to get } 3 + 37 = 40,$$
$$12 + 16 \text{ to get 28,}$$
$$12 + 12 + 12 \text{ to get } 3 + 31 = 34,$$

with 2′ left over from the last combination; thus we would buy

5 16 ft. lengths,
4 12 ft. lengths,
2 8 ft. lengths.

It is now desired to place a stake at each corner and at each joint, and some for support. Let us assume we will place the stakes as shown

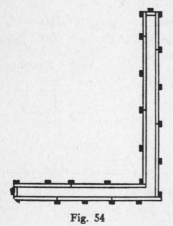

in Figure 54. Accordingly we need 23 stakes. If we make each stake 2 ft. long, we need $(23)(2) = 46$ ft. of 2×2. If we can buy them in 8 ft. lengths, we need

$$\tfrac{46}{8} = 5.75 \quad \text{or} \quad 6 \text{ pieces.}$$

Let us assume that the total cost of the lumber is $14.78 and that we buy $0.75 worth of nails. Normally we would already own a hammer and a saw and would not need to buy them. However, we may not have a trowel, and the cost of one is $3.25. Thus, the total cost of material is

Fig. 54

Concrete	$36.88
Lumber	14.78
Nails	.75
Tools	3.25
Total	$55.66

Repairs and improvements are not regular expenditures, but a certain amount should be allowed for during the year.

F. *Garden and Landscaping*. The external appearance of a home is dependent upon the shrubs, trees, lawn, flowers, etc. In maintaining a well-landscaped place we need to periodically weed and feed the lawn, change flower beds, and replace shrubs. As an example, let us consider the problem of fertilizing the lawn. Suppose our house is located on our lot as shown in Figure 55. Let us assume that all of the lot except the house, driveway, and sidewalk is in lawn. We wish to calculate the area of the lawn.

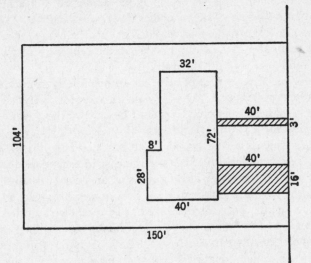

Fig. 55

The total area of the lot is

$$(104)(150) = 15,600 \text{ sq. ft.}$$

The area of the sidewalk is

$$(40)(3) = 120 \text{ sq. ft.}$$

The area of the driveway is

$$(40)(16) = 640 \text{ sq. ft.}$$

The area of the house is

$$(32)(72) + (28)(8) = 2528 \text{ sq. ft.}$$

The area of the lawn is

$$15,600 - 120 - 640 - 2528 = 12,312 \text{ sq. ft.}$$

The garden shop sells fertilizer in bags which are sized to cover a certain area. Let us assume the cost is \$5.98 for a bag which covers

5000 sq. ft. and $3.98 for a bag which covers 2500 sq. ft.   We then need to buy

$$2 \text{ bags at } \$5.98 = \$11.96$$
$$1 \text{ bag at } \$3.98 = 3.98$$
$$\text{Total Cost} = \overline{\$15.94}$$

There will be other expenses connected with the ownership of a house.   The above are representative of the recurring types.

Automobile Ownership

The ownership of a car is becoming more and more of a necessity rather than a luxury.   Many families find it necessary to own more than one car.   We shall now consider the mathematics involved in the ownership of a car.   The expenses can be broken into two parts: (1) the initial cost and (2) the operating cost.

A. *Purchase Price.*   The price of an automobile is usually quoted in the advertisements as "f.o.b.," which stands for *freight on board* and means that the freight charges must be paid by the buyer.   There are also federal and state taxes, handling charges, and charges for accessories which may be added.   The practice today is to itemize these charges, and the buyer can see the cost of each item and make his selection.   If the buyer already owns a car, he can usually "trade it in" on the purchase of a new car.   The automobile sales agency will then make an allowance for the old car.   In recent years this has become a matter of "horse trading" and a buyer usually shops around to see where he can get the best "deal."   A typical situation (at current prices, which are always changing) may be the following:

Price (including Federal Tax)	$2175.00
Automatic Transmission	140.00
Radio	75.00
Heater	85.00
Power Steering	95.00
Power Brakes	45.00
Accessory Group "A"	65.00
Undercoating	20.00
Car plus Accessories	$2700.00
State Tax (3%)	81.00
Freight Charges	65.00
Total Cost	$2846.00
Trade-In Allowance	946.00
Balance Due	$1900.00

Because of the large sum of money which is involved in a car purchasing transaction, many people buy a car on time. Thus in the above listing $1900 could be borrowed from a loan company or bank at a stated rate of interest. There are also finance companies or some automobile dealers who provide for "time buying" and will quote their prices at so much down and so much per month. The maximum carrying time for automobiles is usually three years (36 months), and the charges on a new car are calculated on that basis. Let us assume that we could buy the above listed car with our trade-in as the down payment and the balance to be paid at the rate of $62.28 per month for 36 months. We would then pay

$$(\$62.28)(36) = \$2242.08,$$

and the carrying charges would be

$$\$2242.08 - \$1900.00 = \$342.08,$$

which is (36 months = 3 years)

$$\frac{342.08}{(1900)(3)} = .06 = 6\%.$$

This would make the cost of the automobile

$$\$946.00 + \$2242.08 = \$3188.08.$$

B. *Operating Expenses.* It costs money to operate a car. Let us consider some of the major items.

(a) *License.* All states require that the owner of an automobile have a license to operate the car. The license is sold on an annual basis and varies from state to state. A typical fee is $14.00 per year. Some states include the personal property tax in the license fee, and then the fee may run as high as $150.00 per year or more. Each driver is also required to have a driver's license, which is renewed periodically. The typical fee averages around $2.00 per year.

(b) *Personal Property Tax.* In those states which have a personal property tax, the automobile is a large item. The car is assessed according to its make, size, type, and age, and the tax is calculated according to the annual tax rate. Consequently, this item varies even within a state as the tax rate varies from township to township.

(c) *Insurance.* Most states require that each car owner carry at least a public liability insurance. Responsible car owners desire to have insurance against public liability, property damage, fire, theft, collision, and medical expenses. The rates vary according to the type of car, protection desired, and the location. In many localities at the current time a policy will cost about $100 per year.

(d) *Gas and Oil.* The daily expense of operating a car is mainly concerned with the consumption of gas and oil. This depends upon the car, its efficiency, and the number of miles that the car travels. Most car owners rate their car according to the number of miles per gallon of gas. At the present time most big cars will average around 15 mi./gal., compact cars around 25 mi./gal., and the smaller cars may get into the thirties in their miles per gallon. The price of gasoline varies not only from locality to locality but also from time to time during the year.

(e) *Periodic Service.* In order to keep a car at top efficiency it is recommended that periodic service be performed. Such service includes oil changes, lubrication, and maintenance of transmissions, filters, brakes, etc. Each car manufacturer provides an operating book which will specify the recommended periodic checks.

(f) *Repairs and Replacements.* As the car increases in age certain parts will wear out and fail. The most common items are tires, brakes, batteries, pumps, filters, etc.

(g) *Depreciation.* A big expense is the depreciation of a car, which is primarily due to the manufacturer's practice of changing models each year. The largest depreciation takes place during the first year and may amount to as much as 40% of the purchase price. However, the rate of depreciation diminishes as the years progress, and it is best to calculate the depreciation on an average. Suppose for example that we kept the car which cost us $2846.00 for 4 years and then traded it in on a new one. If we were allowed $1046.00 as the trade-in value, the depreciation would be

$$\$2846 - \$1046 = \$1800$$

or

$$\frac{\$1800}{4} = \$450.00 \text{ per year.}$$

(h) *Miscellaneous.* There are, of course, other expenses such as parking fees, washing, polishing and waxing, etc. Some of these depend upon the habits of the owner. One expense, which is usually viewed as an indirect expense and frequently forgotten altogether, is the result of losing interest because of the investment in the car. Thus if we had not invested the $2846.00 in a car we could take that amount and invest it so that it would return an interest to us. Let us say for example that we had placed it in a bank at 3% compounded semiannually. Over a four year period we would have realized a profit of

$$(\$2846.00)(1.1265) - \$2846.00 = \$360.02.$$

This is not all clear profit as we would have to pay income tax on the interest gained each year. On the other hand, if we had not owned a car, we would have had to pay for public transportation during this time. Most people ignore the loss due to investment potential of the money tied up in a car.

Let us perform a rough cost analysis on the ownership of a car for a period of 4 years, assuming that we traveled 52,000 miles during that time and averaged 15 miles to the gallon. We kept a record of our repairs and services and will estimate some of the miscellaneous expenses. Let us use the purchase price of $2846.00 and the trade-in allowance of $1046.

License (4 × $14.00)	$ 56.00
Gasoline (15 mi./gal, @ 29.9¢)	1036.53
Oil (not including changes, 20 qt. @ 65¢)	13.00
Tires (8 @ $27.50)	220.00
Servicing	312.00
Insurance (4 × $103.00)	412.00
Repairs	96.00
Depreciation ($2846 − $1046)	1800.00
Property Tax	216.00
Miscellaneous	56.00
4 Year Total	$4217.53
Annual Average (÷4)	$1054.38

Insurance

We have already mentioned insurance for specific purposes in the previous sections, such as insurance for the automobile, the house, hospitalization, etc. The philosophy of insurance is to pay small amounts periodically in order to prevent having to pay a large amount at one time or as a form of savings for a future date. The payments made by the insured are called *premiums*, and the amount of the premiums are based upon statistical data which are collected and analyzed by the insurance companies. These analyses are based on higher mathematics and form a very solid foundation for estimating the risks that the insurance company takes. Practically all insurance companies today are founded on sound business principles and are seldom subject to failures. By investing the money collected through the premiums, the company has sufficient profit to pay for its operating expenses and can usually pay dividends to its policy holders.

Premiums and terms of a policy vary from company to company, and the intelligent approach to insurance is to consult an insurance agent who represents a well-established company. These agents have complete brochures which explain their policy and will then calculate the rates which fit the individual's needs. It is always advantageous to consult more than one agent in order to have a basis of comparison.

Much of the insurance needed today can be obtained through group policies, many of which are handled with the co-operation of the place of employment. Policies covering hospitalization, accident, and life can now be obtained in this manner. Automobile insurance is frequently handled by the company which finances the car.

We shall discuss two forms of insurance as examples, but would like to emphasize again that the best information on current insurance practices is obtained from a reputable agent.

A. *Life Insurance.* There are many types of life insurance policies but we can generalize to three kinds:

(1) *Whole-life policy.* The insured pays a fixed premium each year for as long as he lives. The benefit is payable upon death.
(2) *Limited-payment policy.* The insured pays a fixed premium each year for a limited number of years. If he is still living at the end of the specified time, the policy is "paid-up" and no more premiums are paid. The benefit is payable upon death.
(3) *Endowment policy.* The insured pays a fixed premium each year for a limited number of years. The face value of the policy is payable at death or, if the insured is still living at the end of the endowment period, the amount of the benefit is payable to the insured instead of the beneficiary. The payment can either be a lump sum payment or may be paid at so much per month as long as the insured lives. If monthly payments are chosen and there is money remaining in the policy when the insured dies, the balance is paid to the beneficiary.

Premiums for life insurance depend upon the type of policy and the age of the insured. Frequently premiums may be paid annually, semiannually, quarterly, or even monthly. The rate, however, varies and is usually smaller if paid annually.

Example

The premiums for a 20-payment life policy of $1000 for a man at age 20 are

Annually $24.36	Semiannually $12.52	Quarterly $6.52	Monthly $2.30

Compare the annual cost.

Solution. The annual costs are

Annual Rate	$24.36
Semiannual Rate (12.52)(2)	$25.04
Quarterly Rate (6.52)(4)	$26.08
Monthly Rate (2.30)(12)	$27.60

Endowment policies have increased in popularity in recent years as they have the further benefit of providing additional retirement income since the endowment period usually runs until the insured is around 60 years of age.

B. *Property Insurance.* This type of insurance is for the protection of property owners against damages due to hazards of fire, theft, acts of the elements, and public liability. As we have mentioned before, insurance companies now offer "package deals" in which house, personal property, and liability can all be grouped together. The premiums are dependent upon the type of house and its locality. Policies can be written for one, two, three, four, or five year terms and the rate for additional years is usually less than the first year.

Example

Property insurance for a house is written for a five year term at 42¢ a $100 for the first year and 80% of the first premium for each succeeding year. What would be the annual premiums for $20,000.00 worth of insurance?

Solution.

First year: $\left(\dfrac{20,000}{100}\right)(.42) = \$84.00.$

Next 4 Years: ($84.00)(.80) = $67.20 per year.

Investments

It is to the advantage of individuals to invest their surplus cash and thus increase their worth. There are three common ways in which money can be invested: (1) savings accounts in banks; (2) purchase and sale of stocks; and (3) purchase and sale of bonds. We have already discussed savings accounts in Chapter IX. Thus we shall limit our discussion at this time to stocks and bonds.

Stocks and bonds are bought and sold through a broker. Thus the first step is to get associated with a brokerage firm by opening an account with them. For his services the broker will charge a fee both for selling and buying either stocks or bonds. The fee is usually a fixed amount for "lot sale." A lot may be set at 100 shares of stock or a certain number of bonds. Broken lots can be bought or sold, and then the fees are set by the broker.

A. *Bonds.* A bond is a method for a company to borrow money to increase its working capital. Bonds are issued for a definite period of time, at a specified rate of interest, and for a face value of each bond. Thus we could have a 20 year bond at 5% issued at $500 face value. This means that 20 years from date of issue the company promises to pay the holder of the bond $500. In the meantime they will pay 5% of the face value annually in interest. The interest is usually collected twice a year and is obtained by sending *coupons*, which are attached to the bond, to the treasurer of the company for collection. A coupon for the $500 bond at 5% would specify the date due and the amount ($500)(.025) = $12.50 if collected twice a year.

The date when the principal is due is called the *date of maturity*, and the number of years from date of issue to date of maturity is called the *term* of the bond. The face value of the bond is called its *par value.* Although the par value establishes the amount which will be collected at maturity, the price of the bonds may vary according to the demand for them. The price paid is called the *market value;* if it is higher than the par value, the bond is said to be sold at a *premium;* if it is lower, it is said to be sold at a *discount.*

Three things are to be considered when buying bonds: (1) market price, (2) brokerage fee, and (3) the accrued interest. The first two are self-explanatory. The accrued interest enters because the bonds are drawing interest every day even though that interest is collected only twice a year. Let us consider a $1000 bond at 4% with interest dates set at March 1 and September 1. If we purchased the bond on July 8, the previous owner is entitled to the interest from March 1 to July 8. This interest is usually computed on the basis of 360 days to the year. Since there are 129 days between March 1 and July 8, we have

$$1000(\tfrac{129}{360})(.02) = \$7.17$$

in accrued interest. This interest is added to the price of the bond. Of course, this interest plus the balance of the interest from March 1 to September 1 is collected by the new owner on September 1.

In order to determine the profit and rate of income from a bond held to maturity, we calculate the purchase cost without the accrued interest and use this figure as our investment.

Example

We purchase 3 bonds at par value of $1000 for $1047.50 per bond. The interest rate is 5% payable on March 1 and September 1, and the bonds have a maturity date of September 1, 1970. The purchase is made on June 29, 1960. What are our total profit, average annual profit, and average annual rate if the brokerage fee is $3.00 per bond?

Solution.

The market value: (3)(1047.50) = $3142.50
Brokerage Fee: (3)(3.00)   =   9.00
   Our investment: = $3151.50.

Accrued interest:
$$(3)(1000)(\tfrac{120}{360})(.025) = \$25.00.$$

Money collected at maturity: (3)(1000) = $3000.

Profit from interest:

On March 1 and September 1 we collect each time (3)($25) = $75; however, from the amount collected on Sept. 1, 1960 we must subtract the accrued interest of $25 which we paid the previous owner. Thus:

Sept. 1, 1960: $75.00 − $25.00 = $ 50.00
Balance: 9($150.00)   = 1350.00

   Total interest: = 1400.00.

Total money returned:
   $3000.00 + $1400.00   = $4400.00.

Total Profit:
   $4400.00 − $3151.50   = $1248.50.

Elapsed time is $9\tfrac{1}{3}$ years.

Average annual profit:
   $1248.50 ÷ $9\tfrac{1}{3}$ = $ 133.77.

Average annual rate:
$$\frac{133.77}{3151.50} = .0424 \quad \text{or} \quad 4.2\%.$$

B. *Stocks.* When a buyer purchases stocks he is actually buying a share of the ownership of the company's assets. Stocks are issued by the company as shares, of which there are two kinds, namely, *preferred* and *common*. The preferred stock entitles the owner to a fixed per cent of dividend providing the company makes the necessary profit. If after paying the dividend on the preferred stock, there is still some profit remaining, this is then divided among the holders of common stock. Thus the dividend on preferred stock is quoted as a per cent and the dividend on the common stock is quoted at so much per share.

The *par value* of a share of stock is stated on the stock certificate and may be of any amount; many common stock shares have no quoted par value. The *market value* of a stock is the current price at which it is being bought or sold. Dividends are declared according to the profits of the company and may be paid annually, semiannually, quarterly, or monthly.

The market value of stocks which are listed on the stock exchanges are published in most newspapers at the end of a day. Up-to-the-minute prices are obtained from the broker. A typical newspaper listing is shown in the following abbreviated table:

	Sales in 100 s	High	Low	3:00	Net Chg
Dana Corp.	3	34	$33\frac{5}{8}$	34	$+\frac{3}{4}$
Dayco	8	$17\frac{1}{8}$	$16\frac{7}{8}$	17	$-\frac{1}{8}$
Daystrom	34	$36\frac{1}{2}$	35	$36\frac{1}{4}$	$-1\frac{3}{8}$

The first column identifies the stock, the second column shows the sales of the day in 100 shares, the third column gives the high cost for the day, the fourth column the low cost for the day, the fifth column shows the closing price at 3:00 P.M., and the last column shows the net change from yesterday's quotation. Thus in the above listing for Daystrom, there were 3400 shares sold, the high was $36.50 a share, the low $35.00, the closing price $36.25, and this shows a drop of $1.375 a share from yesterday. The quotations are made to the nearest eighth and are usually referred to as "points."

As in the case of bonds, stocks are bought and sold through a broker, who charges a fee for his services. These fees are established for *full lots*, which consist of 100 shares and usually depend upon the value of the stock. Less than 100 shares can be bought and sold, and such transactions are called *odd lots*. Each broker will furnish a list of his fees.

Examples

(2) Mr. Jones bought 300 shares of General Motors common at $43\frac{1}{2}$. The brokerage fee for this value stock is $18 per lot. How much did Mr. Jones pay for the stock?

Solution.

Market price: ($43.50)(300) = $13,050.00
Brokerage fee: ($18)(3)   =   54.00
            Total cost: $13,104.00 *Ans.*

(b) Mr. Smith bought 200 shares of Food Mart at $12\frac{3}{4}$ and sold it at $12\frac{1}{2}$. How much did he lose?

Solution.

	Cost	Return
Market value Brokerage fee	($12.75)(200) = $2550.00	(12.50)(200) = $2500.00
	($15)(2) = 30.00	(15)(2) = 30.00
Totals	$2580.00	$2470.00

In the case of the cost the brokerage fee is added; in the returns it is subtracted. We then have

Cost: $2580.00
Returns: $2470.00
Loss: $ 110.00 *Ans.*

If, in Example (b) above, Mr. Smith held the stock at a time when Food Mart paid a dividend, he would collect this dividend and thus his loss may not have been as large as shown in the example.

Although gains can be made through wise purchasing and selling of stocks, the return on the investment is primarily through the dividends. The rate of return can easily be calculated by dividing the dividend by the total cost of the stocks.

Example

Mr. Jones bought 100 shares at $19\frac{5}{8}$ and paid a brokerage fee of $12.00. The stock paid an annual dividend of $1 a share. What was the rate of return on the investment?

Solution.

Cost: ($19.625)(100) + $12.00 = $1974.50.
Dividend: ($1.00)(100) = $100.00.
Rate: $\dfrac{100.00}{1974.50}$ = .0506 or 5.1%. *Ans.*

Records and Budgets

The administration of a home and family is a small business, and sound business concerns keep records and budgets. It is recommended that all families keep records. Such records need not be elaborate nor require a great deal of time. A systematic approach will conserve both time and energy. Records of income are essential for tax purposes and planned spending. If the income stems from wages, the company paying the salary will usually furnish a statement with each pay check; these figures can then be transferred to a record book.

The amount of bookkeeping necessary varies for each individual, usually depending upon the characteristics of the people involved. Some people prefer to keep detailed records; others, general records; and some, no records at all. We shall illustrate a general scheme. We shall first divide the records into two main categories: (1) Income and (2) Expenses.

A. *Income*. Record of the income can be reduced to monthly entries and calculated on a record sheet in the following manner:

Income — 1960					
Month	Salary	Interest	Dividends	Misc.	Total
Jan.	845.00				845.00
Feb.	845.00			54.00	899.00
Mar.	845.00		108.00	800.00	1753.00
Apr.	845.00				845.00
May	845.00				845.00
June	845.00	56.76	108.00		1009.76
July	920.00				920.00
Aug.	920.00			110.00	1030.00
Sept.	920.00		108.00		1028.00
Oct.	920.00				920.00
Nov.	920.00				920.00
Dec.	920.00	62.84	108.00		1090.84
Totals	10590.00	119.60	432.00	964.00	12105.60

B. *Expenses*. Records of expenses can be grouped in many ways. Some expenses will require daily entries, others can be recorded on a monthly basis. It is desirable to have a monthly summary sheet and individual records of the items which require a further breakdown. Thus the operating expenses connected with the car can be recorded in a book kept in the glove compartment and then summarized each month. The daily cash expenses for the household can be recorded on a work sheet kept in the kitchen. Let us divide the expenses into 10 classes.

1. Food — the grocery bill.
2. Shelter — rent or payments on the mortgage.
3. Car — gas, oil, repairs, license, payments.
4. Taxes — federal income, state income, real estate, personal property.
5. Household — electricity, telephone, gas, fuel, laundry, maintenance, services.
6. Furnishings and Clothing — carpets, drapes, furniture, appliances, utensils, clothing.
7. Health and Education — doctor, dentist, books, magazines, supplies.
8. Insurance — life, accident, hospitalization, house, furnishings, car, etc.
9. Recreation — hobbies, vacation, entertainment.
10. Miscellaneous — if you cannot find any other place for it, put it here.

The form of each work sheet will differ for the items and the details desired. We shall illustrate by a few examples.

Household — Month of January

Date	Item	Amount
1/3	Electricity	$18.32
1/3	Telephone	8.45
1/7	Fuel Oil	36.12
1/10	Gas	9.08
1/10	Repair of Washer	8.75
1/18	Maid Service	8.00
1/20	Plumber Service	6.35
	Total	$95.07

Health and Education — Month of February

Date	Item	Amount
2/7	Dentist (Jack)	$ 5.00
2/8	Books (Helen)	12.25
2/10	School Supplies (Jack)	1.65
2/23	Drug Store	6.28
	Total	$25.18

A summary sheet for the year can then take the form shown in the table below.

EXPENSES — 1960

Month	Food	Shelter	Car	Taxes	Household	F&C	H&E	Ins.	Rec.	Misc.	Total
Jan.	136.42	112.34	18.75	158.85	95.07	4.98	36.12	8.35	6.00	3.25	580.13
Feb.	145.75	112.34	16.33	308.85	88.14	10.42	25.18	111.47	2.00	0.00	820.48
Mar.	123.25	112.34	22.14	158.85	103.69	0.00	4.10	8.35	3.00	4.19	539.91
Apr.	124.68	112.34	33.16	412.42	101.32	9.81	0.00	51.35	30.00	0.00	875.08
May	143.59	112.34	18.75	551.00	73.12	112.00	3.18	127.03	7.50	0.00	1148.51
June	121.12	112.34	21.32	158.85	68.13	98.00	6.72	8.35	12.50	1.36	608.69
July	103.04	112.34	22.48	158.85	56.78	46.15	8.39	8.35	25.00	0.00	541.38
Aug.	162.31	112.34	185.12	158.85	22.18	50.00	0.00	120.88	460.00	9.00	1280.68
Sept.	146.32	112.34	38.16	158.85	108.42	342.68	46.25	8.35	0.00	1.25	962.62
Oct.	142.18	112.34	18.12	551.00	93.56	93.12	32.14	51.35	4.00	1.50	1099.31
Nov.	166.24	112.34	32.14	158.85	102.14	0.00	18.20	8.35	12.00	3.00	613.26
Dec.	173.12	112.34	10.31	158.85	112.69	268.42	104.12	8.35	38.00	9.00	995.20
Totals	1688.02	1348.08	436.78	3094.07	1025.24	1035.58	284.40	520.53	600.00	32.55	10065.25

In the illustration that we have been using we notice that for the year 1960 our total income was $12,105.60 and our expenses amounted to $10,065.25 so that we had a savings of

$$\$12,105.60 - \$10,065.25 = \$2,040.35.$$

This can be considered as a profit for the year, and we can use it for additional investments, earmark it for the purchase of a new item such as a car, or use it to maintain a balance in our checking account. We notice that the last item is necessary since a comparison shows that in some months our expenditures exceeded our income for that particular month.

One reason for keeping records is to permit us to analyze our expenditures and budget for the future. One method of analysis is to calculate the distribution of our expenses on a percentage basis using our income as the standard of measurement. In our illustration we divide each total expense by 12105.60 to obtain the following distribution:

Food	— 13.9%	Furnishings & Clothing	— 8.6%
Shelter	— 11.1%	Health & Education	— 2.3%
Car	— 3.6%	Insurance	— 4.3%
Taxes	— 25.6%	Recreation	— 5.0%
Household	— 8.5%	Miscellaneous	— .3%

Savings — 16.9%

In preparing the budget for next year we should first allow approximately the same distribution as last year. We should also take note of the fact that our expenditures are close to our base salary income ($10590.00 − $10065.25 = $524.75) and that most of the savings are due to interest, dividends, and miscellaneous income.

If we have records for a number of years, we can study the percentages to see how our living expenses may be changing. We should not study the actual amount of money spent, since the cost of living, taxes, and salaries are continuously changing.

However, since we had a good year this year we may decide to buy a new car using this year's profits (savings) to pay for it. If we also need additional items, we should study this year's expenses to see where we can cut down.

Review Exercise X.

Set up a bookkeeping system for your own particular case and determine the distribution of your income and expenses.

APPENDIX

A. Tables. The use of tables of numbers is becoming more and more important in everyday mathematics. We have already found it convenient to use tables of powers and roots, logarithms, compound interest, etc. There are tables of mortgage payments, time tables, mileage tables, stock market listings, etc. The practical use of tables requires thorough familiarity with the types and how to use them. Most tables are read by the intersection of a row with a column.

Example

Find $\sqrt{34}$ in Table I.

Solution.

No.	Square	Cube	Square Root	Cube Root
¦	¦	¦	↓	¦
¦	¦	¦		¦
¦	¦	¦		¦
34		→	5.831	

$$\sqrt{34} = 5.831. \quad Ans.$$

Some tables display more than one quantity. Thus in the above example Table I gives the square, cube, square root, and cube root of numbers. These tables are usually arranged by having the *argument*, value of the running variable, listed in the extreme left-hand column and then each column headed by the corresponding designation of what is contained in that column. Tables I, II, and III are good examples. Table VII, which gives the weight of people according to height, years, and build is an example of multiple column headings, that is, more than one heading for each column.

Other tables may be limited to only one quantity; for example, the logarithms of numbers, Table IV. Here we can make use of an expanded representation of the argument by placing part of it in the extreme left column and part of it as the heading of subsequent columns. The table is read in the same manner as the others by the intersection of the row and column.

Example

Find the mantissa of log 354.

Solution.

N	0	1	2	3	4	5
'						
'						
'						
35					→ 5490	

Time tables for various means of public transportation also work on the principle of the intersection of a row with a column. Their prime purpose is to list the departure and arrival times. Airlines,

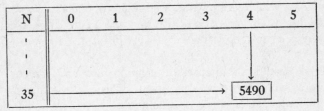

No. 9

WASHINGTON TO CHICAGO | CHICAGO TO WASHINGTON

WASHINGTON (EDT) — NON STOP ONLY — CHICAGO (CDT)

Frequency	Leave	Airports	Arrive	Plane	Airline	Flight No.	Frequency	Leave	Airports	Arrive	Plane	Airline	Flight No.
★Dly	12:59A	M	2:32A	Vis	CA	801TN	★Dly●	7:00A	O	9:35A	DC8	UA	848FT
★XSu	8:00A	M	9:15A	LtE	AA	351	★Dly	8:15A	M	11:42A	Vis	CA	2
★Dly	8:15A	M	9:55A	Vis	CA	1							
★Dly●	8:30A	O	9:15A	DC8	UA	841FT	★XSu	8:50A	M	11:55A	LtE	AA	352
							★Dly	10:45A	M	2:28P	Con	TW	526T
★Dly●	12:00N	O	12:45P	DC8	UA	845FT							
★Dly	12:00N	M	1:45P	Vis	CA	9	★Dly	12:00N	M	3:20P	DC7	UA	722
★Dly	1:00P	M	2:15P	LtE	AA	353	★Dly	12:00N	M	3:30P	Vis	CA	10
★Dly	2:30P	M	4:15P	Vis	CA	15							
							★Dly	1:10P	M	4:15P	LtE	AA	354
★Dly	3:30P	M	4:45P	LtE	AA	357	★Dly	2:00P	M	5:30P	Vis	CA	14
★Dly	4:00P	M	5:45P	Vis	CA	19	★XSa	2:50P	M	5:55P	LtE	AA	356
★XSa	4:45P	M	6:00P	LtE	AA	395							
★Dly	5:00P	M	6:55P	Vis	CA	719T	★Dly	3:25P	O	6:45P	DC7	UA	124T
★Dly	5:30P	M	7:05P	DC7	UA	727	★Dly	4:15P	M	7:45P	Vis	CA	18
							★Dly	5:15P	M	8:45P	Vis	CA	20
★Dly	6:30P	M	8:15P	Vis	CA	21							
★XSa	7:10P	M	8:25P	LtE	AA	355	★Dly	5:40P	M	8:45P	LtE	AA	396
★Dly	8:05P	M	10:10P	Con	TW	525T	★XSa	6:00P	M	9:30P	Vis	CA	22
★XSa	9:00P	M	10:45P	Vis	CA	27	★XSa	6:30P	M	9:35P	LtE	AA	358
★Dly	10:50P	O	12:28A	DC7	UA	129T							
							★Dly	7:45P	M	11:18P	DC6	CA	718T
							★Dly●	9:05P	O	11:40P	DC8	UA	840FT
							★Dly	11:15P	M	2:40A	Vis	CA	800TN

● Serves Washington through Center City Airport

Fig. 56

railroads, buses, and boats all have their own particular terminology, and each table should be studied in order to understand the various abbreviations which are used. Such abbreviations, however, are well explained at the beginning or end of each table. Each company has its own time table; however, in recent years, combined time tables for airline transportation are available between certain of the larger cities. Such tables combine the flights of all the airlines and give the complete run-down of all the flights available between two specified cities. An example is shown in Figure 56, p. 203.

Time tables change periodically and should be checked with the company prior to completing any travel plans.

B. Meter Reading. Advances in technology and science continually provide us with measuring equipment. These are frequently in the form of "meters" such as speedometers, gas meters, barometers, electric meters, water meters, etc. Most meters are in the form of dials with an indicator such as a "needle." The instruments are calibrated against some standard meter and may function because of pressure, temperature, the flow of electricity, or some fluid.

Most people are familiar with the simple meters such as a speed-ometer of the type where the indicator moves across a dial of numbers and if it points to the number "50," the car is traveling 50 miles per hour. Most voltmeters and ammeters work the same way.

Around the house there are three important meters: the gas meter,

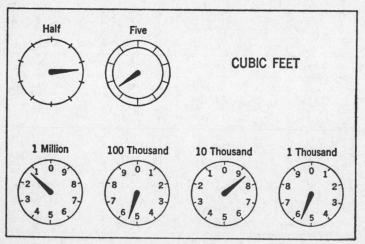

Fig. 57

the electric meter, and the water meter. All operate on the same principle as they measure the flow and consumption. The gas meter measures the number of cubic feet of gas; the electric meter, the number of kilowatt hours; and the water meter, the number of cubic feet. A typical gas meter is shown in Figure 57. This particular meter consists of 6 dials. The top two, measuring the flow of gas down to $\frac{1}{2}$ cu. ft. and 5 cu. ft., rotate the needle in a counterclockwise direction. These two dials are usually not used when reading the meter, as gas rates are quoted at so much per 100 cu. ft. These dials are used to show that the meter is working and calibrated; one complete revolution of the "Half" dial moves the "Five" dial one marker since 10 times $\frac{1}{2}$ equals 5. The bottom four dials give the current meter reading. The extreme right dial is calibrated so that one complete revolution of the needle will measure 1000 cubic feet. Thus each number represents 100 cubic feet. The needle of this dial rotates in a clockwise direction. To the left of this dial we have one revolution representing 10 thousand cubic feet so that each number represents 1000 cu. ft. Here the needle rotates in a counterclockwise direction. The next dial to the left is the 100 thousand dial, and the extreme left dial is the 1 million dial. Notice that the needle of each dial rotates in the opposite direction from that one at its right. The meter is read by taking the number that the needle has just passed. Thus in Figure 57 we read 158,500 cu. ft. The consumption is determined by taking the difference between two consecutive readings. These readings are usually spaced approximately one month apart. Thus if our reading last month was 156,300 cu. ft. and the above reading is now, then we have consumed $158,500 - 156,300 = 2200$ cu. ft. Electric and water meters are very similar in their operation and method of reading except that they usually go down to the unit in their measure.

Some meters such as thermometers and barometers use the fall and rise of a liquid in a calibrated tube. Some recent speedometers use a similar technique with a horizontal colored light reflector on a spring.

C. Fun with Mathematics. There are many interesting situations which occur in mathematics and which can be explained by resorting to advanced mathematical theories. We can, however, amuse ourselves and our friends with the application of the advanced theories without having to study them. In this section we shall present some examples.

I. You are asked to write down the integers from 1 to 9 leaving out the number 8. Thus you write

$$1 \quad 2 \quad 3 \quad 4 \quad 5 \quad 6 \quad 7 \quad 9.$$

You are now asked to choose one integer which you did not write very clearly; some practice in writing it would be in order. Suppose you choose the integer 4. You are now told to multiply the above number by 36, and you get the amazing result

$$\begin{array}{r} 12345679 \\ 36 \\ \hline 74074074 \\ 37037037 \\ \hline 444444444 \end{array}$$

which gives you lots of practice to write the number 4. The choice of the multiplier to obtain a repeated integer in the result is obtained by first choosing the last digit such that when multiplied by 9 the result will be a number which ends in the desired integer; for example, if we want the number to end in 3, we have $(9)(7) = 63$; to end in 5, $(9)(5) = 45$; to end in 7, $(9)(3) = 27$, etc. Having now obtained this number as the last digit of our multiplier we choose the first digit by making the sum of the two digits equal to 9. Thus:

if our second digit is	1	2	3	4	5	6	7	8	9
then our first digit is	8	7	6	5	4	3	2	1	0

II. Try this on your friends.
1. Take your age and double it.
2. Add 5 to the result.
3. Multiply this result by 50.
4. Add the amount of change you have less than $1.
5. Subtract the number of days in a year (365).
6. Tell me the resulting number.
7. Your age is ——; the amount of change you have is ——.

By adding 115 to the number obtained in (6) the age will appear as the first two digits and the amount of change will appear as the last two digits in the final result. Try it.

III. Consider the formula:

$$N = d + 2m + \left[\frac{3(m + 1)}{5}\right] + Y + \left[\frac{Y}{4}\right] - \left[\frac{Y}{100}\right] + \left[\frac{Y}{400}\right] + 2,$$

where:

d is the day of the month;

m is the number of the month in the year, with the qualification that January and February are counted as the 13th and 14th months of the preceding year;

Y is the year;

and the bracket expression means the largest **integer** *not greater than the enclosed number.* Thus $\left[\frac{1958}{4}\right]$ means 489.

If N be divided by 7, the remainder will be the day of the week of a given date where Sunday is counted as the first day, and a remainder of 0 is Saturday.

Example

On what day of the week did January 1, 1961 fall?

Solution. The date January 1, 1961 must be written 13/1/60 according to the definition of m given above. Then we have $d = 1$, $m = 13$, $Y = 1960$, and

$$N = 1 + 26 + \left[\frac{42}{5}\right] + 1960 + \left[\frac{1960}{4}\right] - \left[\frac{1960}{100}\right] + \left[\frac{1960}{400}\right] + 2$$
$$= 1 + 26 + 8 + 1960 + 490 - 19 + 4 + 2$$
$$= 2472$$
$$= (353)(7) + \mathbf{1}.$$

Therefore January 1, 1961 fell on Sunday. Verify the formula for the present date and then calculate the day of the week for your date of birth.

IV. Cryptograms are mathematical puzzles which are concerned with the association of the letters of the alphabet with the digits. The simple cryptogram requires that each letter be replaced by one of the digits 0, 1, 2, 3, 4, 5, 6, 7, 8, 9 and that no digit be represented by more than one letter. Furthermore, the cryptogram should make mathematical sense. The cryptogram

$$\begin{array}{r} send \\ more \\ \hline money \end{array}$$

representing an addition problem can be analyzed in the following manner.

Since the sum of two different digits cannot be greater than 17, $(9 + 8)$, we have $\mathbf{m = 1}$. Thus we have either

$$s + m = s + 1 = o + 10 \quad \text{and} \quad s = o + 9$$

or, since we could have a carry of 1 from $e + o$,

$$s + m + 1 = s + 2 = o + 10 \quad \text{and} \quad s = o + 8.$$

In either case o cannot be greater than 1 for then s would be greater than 9, which is impossible. Now $o \neq 1$ since $m = 1$. Therefore $o = 0$ and $s = 9$ or 8. With $o = 0$ we now have the following four cases for $e + o$:

$$e + o = n \quad \text{or} \quad e + o + 1 = n \quad \text{or} \quad e + o = n + 10$$
$$\text{or} \quad e + o + 1 = n + 10.$$

The first and third of these cannot exist for with $o = 0$ either would have $e = n$, which is impossible. Thus we have

$$e + 1 = n \quad \text{or} \quad e - 9 = n$$

and we have established that there is a carry from the sum of $n + r$ so that

$$n + r = e + 10 \quad \text{or} \quad n + r + 1 = e + 10.$$

If $n = e + 1$, then substituting we have

$$e + 1 + r = e + 10 \quad \text{or} \quad e + 1 + r + 1 = e + 10;$$
$$r = 9 \qquad \text{or} \qquad r = 8.$$

If $n = e - 9$, then

$$e - 9 + r = e + 10 \quad \text{or} \quad e - 9 + r + 1 = e + 10;$$
$$r = 19 \qquad \text{or} \qquad r = 18.$$

But both of these are impossible since r must be a digit. Thus $r = 9$ or 8 and $n \neq e - 9$, and consequently there can be no carry from the sum of $e + o$. If there is no carry from $e + o$, then (since $m = 1$ and $o = 0$)

$$s + m = s + 1 = 10$$

makes $s = 9$ and then $r = 8$ since $r \neq s$. We obtained $r = 8$ from

$$n + r + 1 = e + 10$$

so then the sum of $d + e$ must have a carry and $d + e = y + 10$. We have reduced the problem to the four letters n, d, e, and y, the two equations

$$e + 1 = n \quad \text{and} \quad d + e = y + 10,$$

and the remaining digits 2, 3, 4, 5, 6, 7. Now from the second equation the sum of $d + e$ must be greater than 11 since $y \neq 0 = o$ and $y \neq 1 = m$. Therefore neither d nor e can be 2, 3, or 4 because the sum of any of these with 5, 6, or 7 would not be greater than 11. We are left with 5, 6, or 7 for d and e. Now if $e = 7$, then $n = 1 + 7 = 8 = r$, $\therefore e \neq 7$; if $e = 6$, then d must be 7 to make $y > 11$ and $n = e + 1 = 6 + 1 = 7$ which makes $d = n$, $\therefore e \neq 6$. Consequently $e = 5$ and $n = e + 1 = 6$ so that $d = 7$. With these values

$$e + d = 5 + 7 = 12 = y + 10$$

and $y = 2$. The solution is

0	1	2	5	6	7	8	9
o	*m*	*y*	*e*	*n*	*d*	*r*	*s*

and

$$\begin{array}{r} 9567 \\ \underline{1085} \\ 10652 \quad Ans. \end{array}$$

Problem. What would be the value of the letters to get the largest sum in the cryptogram

$$\begin{array}{r} save \\ \underline{some} \\ money \end{array}$$

if $e = 5$? (*Ans.* The sum is 17650.)

The subject of mathematics furnishes interesting recreation, puzzles, and pastimes, and there are many books written on this phase of mathematics.

HANDY ANDY No. 14

The Earth

Mean diameter = 7912.464 miles
Surface area = 1.967×10^8 square miles
Mass = 1.3173×10^{25} pounds
Mean distance from sun = 92.9 million miles
Velocity of escape = 36,000 ft./sec.
Length of year in days = 365.24219879 − 0.0000000614 $(t - 1900)$
  (t = present year)
Length of day = 23h. 56m. 04.09054s. of mean solar time

The Moon

Mean diameter = 2160 miles
Mass = 1.62×10^{23} pounds
Surface gravity = 5.31 ft./sec.^{2}
Velocity of escape = 7814.4 ft./sec.
Mean distance from earth = 239,000 miles
Average length of sidereal month = 27.321661 days
Visible surface = 41%
Largest visible crater (Bailly) = 183 miles diameter
Highest visible elevation (Newton) = 29,000 feet

ANSWERS TO EXERCISES

Problems I.

1. (a) 430. (b) 4025. (c) 45250. (d) $53.40. (e) $3071.25.
2. (A) $10.04. (B) $7.37. (C) $9.63.
   (D) $9.51. (E) $4.91. (F) $5.88.
   (a) $7.57. (b) $10.43. (c) $3.90.
   (d) $11.97. (e) $13.47. (f) $47.34.
3. A. 4. $59.15. 5. 511 yds. 6. 251.

Problems II.

1. 592.172. 2. $288.92. 3. 1456.0.
4. 47.91. 5. 225. 6. 11.748.
7. $2.66. 8. 235,885.
9. $1952.37, $1943.18, $1795.50, $1398.16, $1279.64.
10. 1.9792 ft.

Problems III.

1. (a) 224. (b) 2608. (c) 69.80. (d) 209.4.
   (e) 2,574,298. (f) 6770.02. (g) 631.7376. (h) 0.13125.
   (i) 0.006204. (j) $465.00. (k) $2.56185. (l) $1634.16.
2. $27.18. 3. $353.52. 4. $547.81.
5. $125,085,000.00. 6. $2711.00.
7. $0.50. 8. $1289.15. 9. $147.15.
10. 5364. 11. 12,088,800.

Problems IV.

1. (a) 2158. (b) $1304\frac{6}{7}$. (c) 1158.
   (d) $1491\frac{7}{8}$. (e) 8.12. (f) 6453.3.
2. (a) 8.66. (b) 0.22. (c) $103.06.
   (d) 71.68. (e) 2924. (f) 1025.56.
3. $74.31. 4. $8.75. 5. 65 ft.
6. 220 states, 259 sq. mi. 7. 187,063.
8. $2275.85. 9. 27. 10. 89.
11. 47.68 mi./hr. 12. $254.72. 13. $258.00.
14. $4.23. 15. 41 mph.

Problems V.

1. (a) $\frac{4}{5}$.   (b) $\frac{5}{6}$.   (c) $\frac{1}{4}$.   (d) $\frac{7}{3}$.   (e) $\frac{3}{8}$.

2. (a) $\frac{13}{12}$.   (b) $\frac{213}{60}$.   (c) $\frac{85}{72}$.   (d) $\frac{19}{12}$.   (e) $5\frac{5}{8}$.   (f) $18\frac{9}{16}$.

3. (a) $\frac{1}{6}$.   (b) $\frac{1}{6}$.   (c) $\frac{79}{363}$.   (d) $\frac{110}{9}$.   (e) $4\frac{1}{15}$.   (f) $6\frac{1}{16}$.

4. (a) $\frac{6}{15}$.   (b) $\frac{9}{8}$.   (c) $\frac{5}{27}$.   (d) $\frac{7}{160}$.

    (e) $\frac{2}{5}$.   (f) $8\frac{1}{6}$.   (g) $70\frac{14}{15}$.   (h) 2.

5. (a) $\frac{3}{2}$.   (b) $\frac{7}{3}$.   (c) 1.   (d) $4\frac{1}{3}$.   (e) $2\frac{1}{4}$.   (f) $4\frac{37}{120}$.

6. (a) $\frac{7}{2}$.   (b) $\frac{1}{4}$.   (c) 3.   (d) $\frac{1}{6}$.

7. 24 inches.   8. Increased by $\frac{14}{30}$ or $\frac{7}{15}$.

9. 15.75 gal., 5.25 gal.   10. 1146.6 mi.

11. $1\frac{9}{16}$ mi.   12. $8\frac{7}{24}$ cups.

13. \$17.17, \$83.38.

Problems VI.

1. (a) .9.   (b) $-.25$.   (c) $-.72$.   (d) $-\frac{3}{56}$.

2. (a) -20.28.   (b) $\frac{1}{9}$.   (c) .005609.

3. (a) -1.802.   (b) $\frac{7}{10}$.   (c) -64.

4. $\frac{11}{6}$.

Review Exercises I.

1. (a) 164.146.   (b) \$46.67.   (c) 46.30.   (d) $\frac{73}{24}$.   (e) $\frac{85}{168}$.

2. (a) 144.890592.   (b) -92.4213.   (c) $-\frac{1}{20}$.   (d) -3.046.

3. On the landing.   4. \$780.00.   5. 18¢.

6. 94¢.   7. No. \$20.65.   8. \$1761.99.

9. 54¢.   10. 9 weeks.   11. 1°.

12. (a) Symmetrical pyramid of digits from 1 to 9.

    (b) Pyramid of 8's.

13. 281.25 lbs.   14. 97.5 mph.

15. \$4.69, \$1.47 tax, \$3.22 to dealer.

Problems VII.

1. 16, 144, 289, 182.25, $\frac{1}{4}$, $\frac{4}{9}$.

2. 27, 729, 5832, 1560.896, $\frac{27}{64}$.

3. 2, 1.587; 2.828, 2; 4.123, 2.571; 9.110, 4.362; 7.483, 3.826.

4. (a) x^5.   (b) $a^3 x^5 y$.   (c) $\frac{1}{2}$.   (d) x^2.

5. (a) (i) 81 sq. ft.;   (ii) 144 sq. ft.;   (iii) 132.25 sq. ft.;

    (iv) x^2 sq. ft.

    (b) (i) 8 ft. by 8 ft.   (ii) 13′ by 13′.   (iii) a ft. by a ft.

Problems VIII.

1. $A = 6$.   2. $P + I = 575$.   3. $d = 64$; $t = 4$.

4. 148.24 mi.   5. $44\frac{4}{9}°$.   6. 56,250.

7. $42\frac{6}{7}$ ohms.   8. $r = \frac{1}{200}$, 100 hours.

Problems IX.

   1. (a) 6.   **(b)** 14.   (c) 25.   (d) 5.   (e) $-\frac{8}{3}a$.   (f) $a - 4$.
   2. 100°, 0°, 37.78°.

Problems X.

   1. $2.75.   **2.** 5.   **3.** $7.02.   **4.** 42, 44, 46.
   5. 10 nickels, 5 quarters, 12 dimes, 7 half-dollars.
   6. 3 hours.   **7.** 3 weeks.   **8.** Coat: $60; dress: $38.
   9. 6.   **10.** 52°, 38°.
   11. 20 acres.   **12.** 79 cents, $1.18.
   13. 320.   **14.** 33.1 cc., 50.3 cc.
   15. $3\frac{3}{11}$ ft., $8\frac{8}{11}$ ft.   **16.** $212.00.
   17. $50.00.   **18.** 12, 18.
   19. 20, 10.

Problems XI.

   1. (a) $42a^3x^3$.   (b) $39ax^3y$.   (c) $21a^2bx^4y^2$.
    (d) $-42axy^2$.   (e) $6x^3 - 9x^2 - 4x + 6$.
    (f) $4x^2 + 2x + 6xy + 3y$.
    (g) $x^2 - 9$.   (h) $4x^2 - 9y^2$.
    (i) $6x^4 + x^3y - 8x^2y^2 + 4xy^3 - 3y^4$.
   2. (a) $9a$.   (b) $3x^2y$.   (c) $-3ax^2$.   (d) $-5x^2y$.
    (e) $4x^2 - 3xy + 2y^2$.

Problems XII.

   1. (a) $(x + y)(x - y)$.   (b) $3ax(1 + y)$.
    (c) $(x - 3)(x + 2)$.   (d) $3a(x + 1)^2$.
    (e) $3(x + y)(x^2 - xy + y^2)$.   (f) $(x + 9y^2)(x - 3y)(x + 3y)$.
    (g) $(2x + 3)(x - 4)$.   (h) $9(2x - 3)(x + 4)$.

Problems XIII.

   1. ± 5.   **2.** 4, 3.   **3.** 2, 5.   **4.** $\frac{3}{2}, \frac{2}{3}$.
   5. $\frac{5}{3}, -3$.   **6.** $-\frac{1}{2} \pm \frac{\sqrt{3}}{2} i$.   **7.** $-\frac{2}{3}, \frac{1}{2}$.
   8. $2 \pm \sqrt{3}$.   **9.** $\frac{1}{2} \pm \frac{1}{3}i$.   **10.** $-\frac{7}{4}, -\frac{3}{2}$.
   11. $-\frac{2}{3}, -\frac{2}{3}$.   **12.** $\frac{1}{2} \pm \sqrt{3}$.   **13.** $\frac{3}{4} \pm \frac{1}{2}i$.
   14. $-1 \pm i\sqrt{2}$.   **15.** 3, 4.
   16. $7\frac{1}{2}$ sec.   **17.** 30 ohms.
   18. 2 ft./sec.   **19.** $3''$ by $6''$.
   20. $\frac{4}{5}$ or $\frac{5}{4}$.

Problems XIV.

  1. (a) 11.   (b) 77.   (c) 15.   (d) 6.

  2. $1.46.   **3.** .333.   **4.** $\frac{3}{16}$.   **5.** The second.

  6. 0.92.   **7.** 555.6 horsepower.

  8. 6 days.   **9.** 31 gal.

10. $22\frac{2}{3}$ lbs., 34 lbs., $45\frac{1}{3}$ lbs.

11. 30 miles.

Review Exercises II.

  1. (a) $2x + 12$.   (b) $36a$.   (c) $20 - 5y$.

  2. (a) 473.   (b) $8x^8 - 4x^3$.   (c) $\dfrac{2}{x}$.

  3. 4, 2.520; 5.196, 3; 9.899, 4.610; 2.646, 1.913.

  4. (a) 452.   (b) $48.   (c) $350.

  5. (a) 6.   (b) $\frac{1}{2}$.   (c) -1.

  6. (a) $6a^2x^3 - 9ax^2y - 2axy + 3y^2$.

     (b) $12ax - 4a^2x^2 + 6ax^2 - 6a^3x^3 - 2a^2x^3 - 3a^3x^4$.

     (c) $7a^2x^2 - 7ax + 8$ with Remainder = 1.

  7. (a) $4(x + 2y)(x - 2y)$.

     (b) $(3x + 7)(2x - 5)$.

     (c) $a(3x - 4)(x + 1)$.

  8. (a) $-\frac{7}{3}, \frac{5}{2}$.   (b) $-3 \pm \sqrt{2}$.   (c) $2 \pm i$.

  9. 6, 8.

10. .353, .300, .289, .261, .297, .333,

     .286, .250, .286, .182, .250, .300.

Problems XV.

  1. (a) .12.   (b) .675.   (c) .045.   (d) .005.

  2. (a) 15%.   (b) 125%.   (c) 0.4%.   (d) $62\frac{1}{2}$%.

  3. (a) 116.5.   (b) 75.   (c) 91.   (d) 1.276875.

  4. $500.00.   **5.** $72.00.   **6.** 20.4%.

  7. 17.1%.   **8.** 45, 345.

  9. 21%.   **10.** $249.02, 8%.

Problems XVI.

  1. $190.25.   **2.** $8.06.   **3.** $1156.26.

  4. The commission, by $111.75.

  5. $2332.50.

  6. (a) $266.67.   (b) $333.33.   (c) $433.33.

  7. $33\frac{1}{3}$%.   **8.** $11,375.80.

  9. $522.05.   **10.** $42,607.50.

Problems XVII.

1. (a) 47°.   (b) 76° 2′.   (c) 7° 16′ 42″.
2. (a) 180° 0′ 0″.   (b) 8° 16′ 19″.   (c) 110° 56° 30″.
3. 11° 36′ 17″.
4. 40°.   5. 30°, 60°, 90°, $\cdots$, 0°.
6. 15 min., 1 min.   7. 86400°.   8. 30°, 4.
9. $1\frac{1}{3}$.   10. 18°.

Problems XVIII.

1. 32° 18′ and 115° 24′.
2. 35.
3. (a) 18.   (b) $\frac{49}{4}\sqrt{3}$.   (c) 84.
4. 34 miles.
5. 7.0714 by rule of thumb; 7.0711 by long-hand method of finding square roots.
6. Congruent right triangles; area = 30.
7. 7.62 mph.   9. 9 miles.
10. $12\frac{8}{13}$, $7\frac{4}{13}$, $5\frac{1}{13}$.   11. 8, 15, 17.
12. 1, 22.

Problems XIX.

1. $135.00.   2. $112.00.   3. 168 sq. ft.
4. $2s = 1.15(2h)$.   5. 12 by 20 in.
6. 64 sq. in., 11.3 in.   7. 510 sq. ft., 540 sq. ft.
8. 32 ft.   9. 4 sq. ft.

Problems XX.

1. (a) 0.2079.   (b) 0.4384.   (c) 1.6643.
2. (a) 38°.   (b) 36°.   (c) 51°.
3. (a) 62°, 13.2, 14.9. (b) 13, 67°, 23°.   (c) $a = 1 = b$.
4. 25 ft.   5. 219.42 lbs., 204.60 lbs.
6. 30 lbs.   7. 1100.7 ft.

Problems XXII.

1. (a) $C = 50.2656$, $A = 201.0624$.
  (b) $C = 37.6992$, $A = 113.0976$.
  (c) $C = 38.9558$, $A = 120.7631$.
2. (a) $r = 3.5014$, $d = 7.0028$, $A = 38.5154$.
  (b) $r = 2.5465$, $d = 5.0929$, $A = 20.3720$.
  (c) $r = 1.8144$, $d = 3.6287$, $A = 10.3421$.
3. $\frac{4}{3}\pi = 4.1888$.   4. 62.832 sq. in.   5. 24.

6. (a) 15.484293 sq. in.   (b) 302.909376 sq. in.
   (c) 16.237975 sq. in.
7. (a) 2.4168, 2.0646.   (b) 10.2333, 9.6159.   (c) 2.5, 2.1650.
8. 4.3302, 3.5355, 2.9390, 2.5, 2.1692, 1.9132, 1.7100, 1.5450,
   1.2940.
9. 179.07 sq. ft.   10. 6.2832 ft.
11. 14.1372 sq. ft.   12. 122.2929 sq. ft.
13. 81.6816 in.   14. 30.1594 in.
15. 0.05915 sq. mi., 1.1427 mi.

Problems XXIII.

1. (a) 216 cu. in.   (b) 56.5488.   (c) 96.
   (d) 229.8478.   (e) $162\pi = 508.9392$.
2. (a) 50.2656.   (b) 144.7649.   (c) 411.5496.
3. 6.63 ft.   4. 43.9824 sq. ft.
5. 1568 lbs. (784 tons).

Review Exercises III.

1. (a) 8° 55′ 48″.   (b) 219° 4′ 48″.   (c) 16° 39′ 14″.
2. 30.   3. 40 sq. in.
4. .8660, .5000, 1.7321, .5774.   5. $C = 36.4426$, $A = 105.6834$.
6. 649.7131.   7. 313.90625 cu. in.
8. 358.4 sq. ft.   9. 250.15 cu. ft.
10. Let A_1 = area of semicircle on a, A_2 = area of semicircle on b,
    A_3 = area of semicircle on c, and A_4 = area of triangle; then
    area of lunes $= A_1 + A_2 + A_4 - A_3 = \frac{1}{4}$ $(a^2 + b^2 - c^2) + A_4$
    $= A_4$.

Problems XXIV.

1. (a) 57 ft.   (b) 49½ ft.   (c) 10596 ft.   (d) 15¾ ft.
2. 127,182,900 sec.   3. 87,120 sq. ft.
4. 126.98 or 127 barrels.   5. 112 pecks.
6. 139,968 cu. in.   7. 96,000 oz.
8. 6 sq. yds. 4 sq. ft. 4 sq. in.
9. 0.0198 oz.
10. 7 days 14 hrs. 30 min.

Problems XXV.

1. 898.2035 qts., 1496.06 in., 12.86 ft., 439.521368 qts.,
   26.4554676 lbs.
2. 114.71012 cc.   3. 17.36 sq. in.
4. 10.886218 mt.   5. 23,622 in.

Review Exercises IV.

1. (a) 16,023 ft.   (b) 130,680 sq. ft.   (c) $9\frac{3}{4}$ bu.
2. (a) 30.8647122 lbs.   (b) 20.07749 qts.
(c) 8.1771356 cu. in.
3. (a) 93.6466384 sq. m.   (b) 6.881031 cu. m.
(c) 6.607188 liters.
4. $2548.50.   **5.** $453.33.

Problems XXVI.

1. (a) 0.   (b) −2.   (c) 5.   (d) −4.
2. (a) 1.5988.   (b) 9.5393 − 10.   (c) 2.9240.   (d) 1.8375.
3. (a) 495.7.   (b) 0.002221.   (c) 0.7308.

Review Exercises VI.

1. (a) 1.693.   (b) 0.1175.   (c) 1.752.
2. 6859 ft.   **3.** 892.8.   **4.** 4244 sq. in.   **5.** 304.1 cc.
6. 32.22%.   **7.** 1.662 sec.

Review Exercises VII.

1. (a) 504.   (b) 11,793,600.   (c) 32,760.
2. (a) 84.   (b) 376,740.   (c) 1,365.
3. 15.   **4.** 360.   **5.** 924.
6. 59,875,200.   **7.** 10 cents.   **8.** $\frac{1}{2}$, $\frac{1}{26}$.
9. $1.67.   **10.** $\frac{3}{13}$.

Problems XXVII.

1. (a) $128.   (b) $144.   (c) $140.
(d) $12.   (e) $17.60.   (f) $25.67.
2. (a) $1351.44.   (b) $1351.80.   (c) $2167.32.
3. 35 years.

Problems XXVIII.

1. (a) Aug. 4, 1960.   (b) Aug. 19, 1960.   (c) Oct. 3, 1960.
(d) Sept. 5, 1960.   (e) Jan. 5, 1961.   (f) Oct. 5, 1962.
2. $8.12, $491.88.
3. 11.25%.

Review Exercises VIII.

1. (a) $325.   (b) $15.33.   (c) $52.50.
2. (a) $1343.90.   (b) $1126.80.
3. $11, $589.   **4.** 11.73%.
5. $781.20.   **6.** $2710.40.   **7.** $23\frac{1}{2}$ years.

Review Exercises IX.

1. $170.69.
2. Selling Price: $25,750.50   100%
   Net Cost: 11,800.00   45.8%
   Overhead: 2,147.75   8.3%
   Gross Cost: 13,947.75   54.2%
   Margin: 13,950.50   54.2%
   Profit: 11,802.75   45.8%
3. 33%.   4. 20.6%.   5. $756.

TABLES

Tables

Table I.   Powers and Roots.

No.	Square	Cube	Square Root	Cube Root	No.	Square	Cube	Square Root	Cube Root
1	1	1	1.000	1.000	51	2601	132651	7.141	3.708
2	4	8	1.414	1.260	52	2704	140608	7.211	3.733
3	9	27	1.732	1.442	53	2809	148877	7.280	3.756
4	16	64	2.000	1.587	54	2916	157464	7.348	3.780
5	25	125	2.236	1.710	55	3025	166375	7.416	3.803
6	36	216	2.449	1.817	56	3136	175616	7.483	3.826
7	49	343	2.646	1.913	57	3249	185193	7.550	3.849
8	64	512	2.828	2.000	58	3364	195112	7.616	3.871
9	81	729	3.000	2.080	59	3481	205379	7.681	3.893
10	100	1000	3.162	2.154	60	3600	216000	7.746	3.915
11	121	1331	3.317	2.224	61	3721	226981	7.810	3.936
12	144	1728	3.464	2.289	62	3844	238328	7.874	3.958
13	169	2197	3.606	2.351	63	3969	250047	7.937	3.979
14	196	2744	3.742	2.410	64	4096	262144	8.000	4.000
15	225	3375	3.873	2.466	65	4225	274625	8.062	4.021
16	256	4096	4.000	2.520	66	4356	287496	8.124	4.041
17	289	4913	4.123	2.571	67	4489	300763	8.185	4.062
18	324	5832	4.243	2.621	68	4624	314432	8.246	4.082
19	361	6859	4.359	2.668	69	4761	328509	8.307	4.102
20	400	8000	4.472	2.714	70	4900	343000	8.367	4.121
21	441	9261	4.583	2.759	71	5041	357911	8.426	4.141
22	484	10648	4.690	2.802	72	5184	373248	8.485	4.160
23	529	12167	4.796	2.844	73	5329	389017	8.544	4.179
24	576	13824	4.899	2.884	74	5476	405224	8.602	4.198
25	625	15625	5.000	2.924	75	5625	421875	8.660	4.217
26	676	17576	5.099	2.962	76	5776	438976	8.718	4.236
27	729	19683	5.196	3.000	77	5929	456533	8.775	4.254
28	784	21952	5.292	3.037	78	6084	474552	8.832	4.273
29	841	24389	5.385	3.072	79	6241	493039	8.888	4.291
30	900	27000	5.477	3.107	80	6400	512000	8.944	4.309
31	961	29791	5.568	3.141	81	6561	531441	9.000	4.327
32	1024	32768	5.657	3.175	82	6724	551368	9.055	4.344
33	1089	35937	5.745	3.208	83	6889	571787	9.110	4.362
34	1156	39304	5.831	3.240	84	7056	592704	9.165	4.380
35	1225	42875	5.916	3.271	85	7225	614125	9.220	4.397
36	1296	46656	6.000	3.302	86	7396	636056	9.274	4.414
37	1369	50653	6.083	3.332	87	7569	658503	9.327	4.431
38	1444	54872	6.164	3.362	88	7744	681472	9.381	4.448
39	1521	59319	6.245	3.391	89	7921	704969	9.434	4.465
40	1600	64000	6.325	3.420	90	8100	729000	9.487	4.481
41	1681	68921	6.403	3.448	91	8281	753571	9.539	4.498
42	1764	74088	6.481	3.476	92	8464	778688	9.592	4.514
43	1849	79507	6.557	3.503	93	8649	804357	9.644	4.531
44	1936	85184	6.633	3.530	94	8836	830584	9.695	4.547
45	2025	91125	6.708	3.557	95	9025	857375	9.747	4.563
46	2116	97336	6.782	3.583	96	9216	884736	9.798	4.579
47	2209	103823	6.856	3.609	97	9409	912673	9.849	4.595
48	2304	110592	6.928	3.634	98	9604	941192	9.899	4.610
49	2401	117649	7.000	3.659	99	9801	970299	9.950	4.626
50	2500	125000	7.071	3.684	100	10000	1000000	10.000	4.642

Table II. Constants.

Term	Value	Reciprocal	Square	Square Root	Log
$\pi/6$	.52360	1.90986	.27416	.72360	71900
$\pi/4$	.78540	1.27324	.61685	.88623	89509
$\pi/3$	1.04720	0.95493	1.47361	1.02333	02003
$\pi/2$	1.57080	0.63662	2.46740	1.25331	19612
$2\pi/3$	2.09440	0.47746	4.38649	1.44720	32106
$3\pi/4$	2.35619	0.42441	5.55165	1.53499	37221
$5\pi/6$	2.61799	0.38197	6.85389	1.61802	41797
π	3.14159	0.31831	9.86960	1.77245	49715
$7\pi/6$	3.66519	0.27284	13.43363	1.91447	56410
$5\pi/4$	3.92699	0.25465	15.42126	1.98166	59406
$4\pi/3$	4.18879	0.23873	17.54596	2.04665	62209
$3\pi/2$	4.71239	0.21221	22.20661	2.17080	67324
$5\pi/3$	5.23599	0.19099	27.41557	2.28823	71900
$11\pi/6$	5.75959	0.17362	33.17284	2.39991	76039
2π	6.28319	0.15915	39.47842	2.50663	79818
e	2.71828	0.36788	7.38906	1.64872	43429

1 radian = 57.29577 95131 degrees.

1 degree = 0.01745 32925 radians.

Table III. Natural Trigonometric Functions.

Angle °	sin	cos	tan	cot
0	.0000	1.0000	.0000	--
1	.0175	.9998	.0175	57.2900
2	.0349	.9994	.0349	28.6363
3	.0523	.9986	.0524	19.0811
4	.0698	.9976	.0699	14.3007
5	.0872	.9962	.0875	11.4301
6	.1045	.9945	.1051	9.5144
7	.1219	.9925	.1228	8.1443
8	.1392	.9903	.1405	7.1154
9	.1564	.9877	.1584	6.3138
10	.1736	.9848	.1763	5.6713
11	.1908	.9816	.1944	5.1446
12	.2079	.9781	.2126	4.7046
13	.2250	.9744	.2309	4.3315
14	.2419	.9703	.2493	4.0108
15	.2588	.9659	.2679	3.7321
16	.2756	.9613	.2867	3.4874
17	.2924	.9563	.3057	3.2709
18	.3090	.9511	.3249	3.0777
19	.3256	.9455	.3443	2.9042
20	.3420	.9397	.3640	2.7475
21	.3584	.9336	.3839	2.6051
22	.3746	.9272	.4040	2.4751
23	.3907	.9205	.4245	2.3559
24	.4067	.9135	.4452	2.2460
25	.4226	.9063	.4663	2.1445
26	.4384	.8988	.4877	2.0503
27	.4540	.8910	.5095	1.9626
28	.4695	.8829	.5317	1.8807
29	.4848	.8746	.5543	1.8040
30	.5000	.8660	.5774	1.7321
31	.5150	.8572	.6009	1.6643
32	.5299	.8480	.6249	1.6003
33	.5446	.8387	.6494	1.5399
34	.5592	.8290	.6745	1.4826
35	.5736	.8192	.7002	1.4281
36	.5878	.8090	.7265	1.3764
37	.6018	.7986	.7536	1.3270
38	.6157	.7880	.7813	1.2799
39	.6293	.7771	.8098	1.2349
40	.6428	.7660	.8391	1.1918
41	.6561	.7547	.8693	1.1504
42	.6691	.7431	.9004	1.1106
43	.6820	.7314	.9325	1.0724
44	.6947	.7193	.9657	1.0355
45	.7071	.7071	1.0000	1.0000

Table III. Continued.

Angle °	sin	cos	tan	cot
45	.7071	.7071	1.0000	1.0000
46	.7193	.6947	1.0355	.9657
47	.7314	.6820	1.0724	.9325
48	.7431	.6691	1.1106	.9004
49	.7547	.6561	1.1504	.8693
50	.7660	.6428	1.1918	.8391
51	.7771	.6293	1.2349	.8098
52	.7880	.6157	1.2799	.7813
53	.7986	.6018	1.3270	.7536
54	.8090	.5878	1.3764	.7265
55	.8192	.5736	1.4281	.7002
56	.8290	.5592	1.4826	.6745
57	.8387	.5446	1.5399	.6494
58	.8480	.5299	1.6003	.6249
59	.8572	.5150	1.6643	.6009
60	.8660	.5000	1.7321	.5774
61	.8746	.4848	1.8040	.5543
62	.8829	.4695	1.8807	.5317
63	.8910	.4540	1.9626	.5095
64	.8988	.4384	2.0503	.4877
65	.9063	.4226	2.1445	.4663
66	.9135	.4067	2.2460	.4452
67	.9205	.3907	2.3559	.4245
68	.9272	.3746	2.4751	.4040
69	.9336	.3584	2.6051	.3839
70	.9397	.3420	2.7475	.3640
71	.9455	.3256	2.9042	.3443
72	.9511	.3090	3.0777	.3249
73	.9563	.2924	3.2709	.3057
74	.9613	.2756	3.4874	.2867
75	.9659	.2588	3.7321	.2679
76	.9703	.2419	4.0108	.2493
77	.9744	.2250	4.3315	.2309
78	.9781	.2079	4.7046	.2126
79	.9816	.1908	5.1446	.1944
80	.9848	.1736	5.6713	.1763
81	.9877	.1564	6.3138	.1584
82	.9903	.1392	7.1154	.1405
83	.9925	.1219	8.1443	.1228
84	.9945	.1045	9.5144	.1051
85	.9962	.0872	11.4301	.0875
86	.9976	.0698	14.3007	.0699
87	.9986	.0523	19.0811	.0524
88	.9994	.0349	28.6363	.0349
89	.9998	.0175	57.2900	.0175
90	1.0000	.0000		.0000

Tables

Table IV. Four-Place Common Logarithms.

N	0	1	2	3	4	5	6	7	8	9
10	0000	0043	0086	0128	0170	0212	0253	0294	0334	0374
11	0414	0453	0492	0531	0569	0607	0645	0682	0719	0755
12	0792	0828	0864	0899	0934	0969	1004	1038	1072	1106
13	1139	1173	1206	1239	1271	1303	1335	1367	1399	1430
14	1461	1492	1523	1553	1584	1614	1644	1673	1703	1732
15	1761	1790	1818	1847	1875	1903	1931	1959	1987	2014
16	2041	2068	2095	2122	2148	2175	2201	2227	2253	2279
17	2304	2330	2355	2380	2405	2430	2455	2480	2504	2529
18	2553	2577	2601	2625	2648	2672	2695	2718	2742	2765
19	2788	2810	2833	2856	2878	2900	2923	2945	2967	2989
20	3010	3032	3054	3075	3096	3118	3139	3160	3181	3201
21	3222	3243	3263	3284	3304	3324	3345	3365	3385	3404
22	3424	3444	3464	3483	3502	3522	3541	3560	3579	3598
23	3617	3636	3655	3674	3692	3711	3729	3747	3766	3784
24	3802	3820	3838	3856	3874	3892	3909	3927	3945	3962
25	3979	3997	4014	4031	4048	4065	4082	4099	4116	4133
26	4150	4166	4183	4200	4216	4232	4249	4265	4281	4298
27	4314	4330	4346	4362	4378	4393	4409	4425	4440	4456
28	4472	4487	4502	4518	4533	4548	4564	4579	4594	4609
29	4624	4639	4654	4669	4683	4698	4713	4728	4742	4757
30	4771	4786	4800	4814	4829	4843	4857	4871	4886	4900
31	4914	4928	4942	4955	4969	4983	4997	5011	5024	5038
32	5051	5065	5079	5092	5105	5119	5132	5145	5159	5172
33	5185	5198	5211	5224	5237	5250	5263	5276	5289	5302
34	5315	5328	5340	5353	5366	5378	5391	5403	5416	5428
35	5441	5453	5465	5478	5490	5502	5514	5527	5539	5551
36	5563	5575	5587	5599	5611	5623	5635	5647	5658	5670
37	5682	5694	5705	5717	5729	5740	5752	5763	5775	5786
38	5798	5809	5821	5832	5843	5855	5866	5877	5888	5899
39	5911	5922	5933	5944	5955	5966	5977	5988	5999	6010
40	6021	6031	6042	6053	6064	6075	6085	6096	6107	6117
41	6128	6138	6149	6160	6170	6180	6191	6201	6212	6222
42	6232	6243	6253	6263	6274	6284	6294	6304	6314	6325
43	6335	6345	6355	6365	6375	6385	6395	6405	6415	6425
44	6435	6444	6454	6464	6474	6484	6493	6503	6513	6522
45	6532	6542	6551	6561	6571	6580	6590	6599	6609	6618
46	6628	6637	6646	6656	6665	6675	6684	6693	6702	6712
47	6721	6730	6739	6749	6758	6767	6776	6785	6794	6803
48	6812	6821	6830	6839	6848	6857	6866	6875	6884	6893
49	6902	6911	6920	6928	6937	6946	6955	6964	6972	6981
50	6990	6998	7007	7016	7024	7033	7042	7050	7059	7067
51	7076	7084	7093	7101	7110	7118	7126	7135	7143	7152
52	7160	7168	7177	7185	7193	7202	7210	7218	7226	7235
53	7243	7251	7259	7267	7275	7284	7292	7300	7308	7316
54	7324	7332	7340	7348	7356	7364	7372	7380	7388	7396

Table IV. Continued.

N	0	1	2	3	4	5	6	7	8	9
55	7404	7412	7419	7427	7435	7443	7451	7459	7466	7474
56	7482	7490	7497	7505	7513	7520	7528	7536	7543	7551
57	7559	7566	7574	7582	7589	7597	7604	7612	7619	7627
58	7634	7642	7649	7657	7664	7672	7679	7686	7694	7701
59	7709	7716	7723	7731	7738	7745	7752	7760	7767	7774
60	7782	7789	7796	7803	7810	7818	7825	7832	7839	7846
61	7853	7860	7868	7875	7882	7889	7896	7903	7910	7917
62	7924	7931	7938	7945	7952	7959	7966	7973	7980	7987
63	7993	8000	8007	8014	8021	8028	8035	8041	8048	8055
64	8062	8069	8075	8082	8089	8096	8102	8109	8116	8122
65	8129	8136	8142	8149	8156	8162	8169	8176	8182	8189
66	8195	8202	8209	8215	8222	8228	8235	8241	8248	8254
67	8261	8267	8274	8280	8287	8293	8299	8306	8312	8319
68	8325	8331	8338	8344	8351	8357	8363	8370	8376	8382
69	8388	8395	8401	8407	8414	8420	8426	8432	8439	8445
70	8451	8457	8463	8470	8476	8482	8488	8494	8500	8506
71	8513	8519	8525	8531	8537	8543	8549	8555	8561	8567
72	8573	8579	8585	8591	8597	8603	8609	8615	8621	8627
73	8633	8639	8645	8651	8657	8663	8669	8675	8681	8686
74	8692	8698	8704	8710	8716	8722	8727	8733	8739	8745
75	8751	8756	8762	8768	8774	8779	8785	8791	8797	8802
76	8808	8814	8820	8825	8831	8837	8842	8848	8854	8859
77	8865	8871	8876	8882	8887	8893	8899	8904	8910	8915
78	8921	8927	8932	8938	8943	8949	8954	8960	8965	8971
79	8976	8982	8987	8993	8998	9004	9009	9015	9020	9025
80	9031	9036	9042	9047	9053	9058	9063	9069	9074	9079
81	9085	9090	9096	9101	9106	9112	9117	9122	9128	9133
82	9138	9143	9149	9154	9159	9165	9170	9175	9180	9186
83	9191	9196	9201	9206	9212	9217	9222	9227	9232	9238
84	9243	9248	9253	9258	9263	9269	9274	9279	9284	9289
85	9294	9299	9304	9309	9315	9320	9325	9330	9335	9340
86	9345	9350	9355	9360	9365	9370	9375	9380	9385	9390
87	9395	9400	9405	9410	9415	9420	9425	9430	9435	9440
88	9445	9450	9455	9460	9465	9469	9474	9479	9484	9489
89	9494	9499	9504	9509	9513	9518	9523	9528	9533	9538
90	9542	9547	9552	9557	9562	9566	9571	9576	9581	9586
91	9590	9595	9600	9605	9609	9614	9619	9624	9628	9633
92	9638	9643	9647	9652	9657	9661	9666	9671	9675	9680
93	9685	9689	9694	9699	9703	9708	9713	9717	9722	9727
94	9731	9736	9741	9745	9750	9754	9759	9763	9768	9773
95	9777	9782	9786	9791	9795	9800	9805	9809	9814	9818
96	9823	9827	9832	9836	9841	9845	9850	9854	9859	9863
97	9868	9872	9877	9881	9886	9890	9894	9899	9903	9908
98	9912	9917	9921	9926	9930	9934	9939	9943	9948	9952
99	9956	9961	9965	9969	9974	9978	9983	9987	9991	9996

Table V.   Compound Amount of $1.

n	1%	1½%	2%	2½%	3%	3½%
1	1.0100	1.0150	1.0200	1.0250	1.0300	1.0350
2	1.0201	1.0302	1.0404	1.0506	1.0609	1.0712
3	1.0303	1.0457	1.0612	1.0769	1.0927	1.1087
4	1.0406	1.0614	1.0824	1.1038	1.1255	1.1475
5	1.0510	1.0773	1.1041	1.1314	1.1573	1.1877
6	1.0615	1.0934	1.1262	1.1597	1.1941	1.2293
7	1.0721	1.1098	1.1487	1.1887	1.2299	1.2723
8	1.0829	1.1265	1.1717	1.2184	1.2668	1.3168
9	1.0937	1.1434	1.1951	1.2489	1.3048	1.3629
10	1.1046	1.1605	1.2190	1.2801	1.3439	1.4106
11	1.1157	1.1779	1.2434	1.3121	1.3842	1.4600
12	1.1268	1.1956	1.2682	1.3449	1.4258	1.5111
13	1.1381	1.2136	1.2936	1.3785	1.4685	1.5640
14	1.1495	1.2318	1.3195	1.4130	1.5126	1.6187
15	1.1610	1.2502	1.3459	1.4483	1.5580	1.6753
16	1.1726	1.2690	1.3728	1.4845	1.6047	1.7340
17	1.1843	1.2880	1.4002	1.5216	1.6528	1.7947
18	1.1961	1.3073	1.4282	1.5597	1.7024	1.8575
19	1.2081	1.3270	1.4568	1.5987	1.7535	1.9225
20	1.2202	1.3469	1.4859	1.6386	1.8061	1.9898
21	1.2324	1.3671	1.5157	1.6796	1.8603	2.0594
22	1.2447	1.3876	1.5460	1.7216	1.9161	2.1315
23	1.2572	1.4084	1.5769	1.7646	1.9763	2.2061
24	1.2697	1.4295	1.6084	1.8087	2.0328	2.2833
25	1.2824	1.4509	1.6406	1.8539	2.0938	2.2632
26	1.2953	1.4727	1.6734	1.9003	2.1566	2.4460
27	1.3082	1.4948	1.7069	1.9478	2.2213	2.5316
28	1.3213	1.5172	1.7410	1.9965	2.2879	2.6202
29	1.3345	1.5400	1.7758	2.0464	2.3566	2.7119
30	1.3478	1.5631	1.8114	2.0976	2.4273	2.8068
31	1.3613	1.5865	1.8476	2.1500	2.5001	2.9050
32	1.3749	1.6103	1.8845	2.2038	2.5751	3.0067
33	1.3887	1.6345	1.9222	2.2589	2.6523	3.1119
34	1.4026	1.6590	1.9607	2.3153	2.7319	3.2209
35	1.4166	1.6839	1.9999	2.3732	2.8139	3.3336
36	1.4308	1.7091	2.0399	2.4325	2.8983	3.4503
37	1.4451	1.7348	2.0807	2.4933	2.9852	3.5710
38	1.4595	1.7608	2.1223	2.5557	3.0748	3.6960
39	1.4741	1.7872	2.1647	2.6196	3.1670	3.8254
40	1.4889	1.8140	2.2080	2.6851	3.2620	3.9593
41	1.5038	1.8412	2.2522	2.7522	3.3599	4.0978
42	1.5188	1.8688	2.2972	2.8210	3.4607	4.2413
43	1.5340	1.8969	2.3432	2.8915	3.5645	4.3897
44	1.5493	1.9253	2.3901	2.9638	3.6715	4.5433
45	1.5648	1.9542	2.4379	3.0379	3.7816	4.7024
46	1.5805	1.9835	2.4866	3.1139	3.8950	4.8669
47	1.5963	2.0133	2.5363	3.1917	4.0119	5.0373
48	1.6122	2.0435	2.5871	3.2715	4.1323	5.2136
49	1.6283	2.0741	2.6388	3.3533	4.2562	5.3961
50	1.6446	2.1052	2.6916	3.4371	4.3839	5.5849

Table V. Continued.

n	4%	4½%	5%	5½%	6%	6½%
1	1.0400	1.0450	1.0500	1.0550	1.0600	1.0650
2	1.0816	1.0920	1.1025	1.1130	1.1236	1.1342
3	1.1249	1.1412	1.1576	1.1742	1.1910	1.2079
4	1.1699	1.1925	1.2155	1.2388	1.2625	1.2865
5	1.2167	1.2462	1.2763	1.3070	1.3382	1.3701
6	1.2653	1.3023	1.3401	1.3788	1.4185	1.4591
7	1.3159	1.3609	1.4071	1.4547	1.5036	1.5540
8	1.3686	1.4221	1.4755	1.5347	1.5938	1.6550
9	1.4233	1.4861	1.5513	1.6191	1.6895	1.7626
10	1.4802	1.5530	1.6289	1.7081	1.7908	1.8771
11	1.5395	1.6229	1.7103	1.8021	1.8983	1.9992
12	1.6010	1.6959	1.7959	1.9012	2.0122	2.1291
13	1.6651	1.7722	1.8856	2.0058	2.1329	2.2675
14	1.7317	1.8519	1.9799	2.1161	2.2609	2.4149
15	1.8009	1.9353	2.0789	2.2325	2.3966	2.5718
16	1.8730	2.0224	2.1829	2.3553	2.5404	2.7390
17	1.9479	2.1134	2.2920	2.4848	2.6928	2.9170
18	2.0258	2.2085	2.4066	2.6215	2.8543	3.1067
19	2.1068	2.3079	2.5270	2.7656	3.0256	3.3086
20	2.1911	2.4117	2.6533	2.9178	3.2071	3.5236
21	2.2788	2.5202	2.7860	3.0782	3.3996	3.7527
22	2.3699	2.6337	2.9253	3.2475	3.6035	3.9966
23	2.4647	2.7522	3.0715	3.4262	3.8197	4.2564
24	2.5633	2.8760	3.2251	3.6146	4.0489	4.5331
25	2.6658	3.0054	3.3864	3.8134	4.2919	4.8277
26	2.7725	3.1407	3.5557	4.0231	4.5494	5.1415
27	2.8834	3.2820	3.7335	4.2444	4.8223	5.4757
28	2.9987	3.4297	3.9201	4.4778	5.1117	5.8316
29	3.1187	3.5840	4.1161	4.7241	5.4184	6.2107
30	3.2434	3.7453	4.3219	4.9840	5.7435	6.6144
31	3.3731	3.9138	4.5380	5.2581	6.0881	7.0443
32	3.5081	4.0900	4.7649	5.5473	6.4534	7.5022
33	3.6484	4.2740	5.0032	5.8524	6.8406	7.9898
34	3.7943	4.4664	5.2533	6.1742	7.2510	8.5092
35	3.9461	4.6673	5.5160	6.5138	7.6861	9.0623
36	4.1039	4.8774	5.7918	6.8721	8.1473	9.6513
37	4.2681	5.0969	6.0814	7.2501	8.6361	10.2786
38	4.4388	5.3262	6.3855	7.6488	9.1543	10.9467
39	4.6164	5.5659	6.7048	8.0695	9.7035	11.6583
40	4.8010	5.8164	7.0400	8.5133	10.2857	12.4161
41	4.9931	6.0781	7.3920	8.9815	10.9029	13.2231
42	5.1928	6.3516	7.7616	9.4755	11.5570	14.0826
43	5.4005	6.6374	8.1497	9.9967	12.2505	14.9980
44	5.6165	6.9361	8.5572	10.5465	12.9855	15.9729
45	5.8412	7.2482	8.9850	11.1266	13.7646	17.0111
46	6.0748	7.5744	9.4343	11.7385	14.5905	18.1168
47	6.3178	7.9153	9.9060	12.3841	15.4659	19.2944
48	6.5705	8.2715	10.4013	13.0653	16.3939	20.5485
49	6.8333	8.6437	10.9213	13.7838	17.3775	21.8842
50	7.1067	9.0326	11.4674	14.5420	18.4202	23.3067

Table VI.　Present Worth of $1.

n	1%	1½%	2%	2½%	3%	3½%
1	.99010	.98522	.98039	.97561	.97087	.96618
2	.98030	.97066	.96117	.95181	.94260	.93351
3	.97059	.95632	.94232	.92860	.91514	.90194
4	.96098	.94218	.92385	.90595	.88849	.87144
5	.95147	.92826	.90573	.88385	.86261	.84197
6	.94205	.91454	.88797	.86230	.83748	.81350
7	.93272	.90103	.87056	.84127	.81309	.78599
8	.92348	.88771	.85349	.82075	.78941	.75941
9	.91434	.87459	.83676	.80073	.76642	.73373
10	.90529	.86167	.82035	.78120	.74409	.70892
11	.89632	.84893	.80426	.76214	.72242	.68495
12	.88745	.83639	.78849	.74356	.70138	.66178
13	.87866	.82403	.77303	.72542	.68095	.63940
14	.86996	.81185	.75788	.70773	.66112	.61778
15	.86135	.79985	.74301	.69047	.64186	.59689
16	.85282	.78803	.72845	.67362	.62317	.57671
17	.84438	.77639	.71416	.65720	.60502	.55720
18	.83602	.76491	.70016	.64117	.58739	.53836
19	.82774	.75361	.68643	.62553	.57029	.52016
20	.81954	.74247	.67297	.61027	.55368	.50257
21	.81143	.73150	.65978	.59539	.53755	.48557
22	.80340	.72069	.64684	.58086	.52189	.46915
23	.79544	.71004	.63416	.56670	.50669	.45329
24	.78757	.69954	.62172	.55288	.49193	.43796
25	.77977	.68921	.60953	.53939	.47761	.42315
26	.77205	.67902	.59758	.52623	.46369	.40884
27	.76440	.66899	.58586	.51340	.45019	.39501
28	.75684	.64910	.57437	.50088	.43708	.38165
29	.74934	.64936	.56311	.48866	.42435	.36875
30	.74192	.63976	.55207	.47674	.41199	.35628
31	.73458	.63031	.54125	.46511	.39999	.34423
32	.72730	.62099	.53063	.45377	.38834	.33259
33	.72010	.61182	.52023	.44270	.37703	.32134
34	.71297	.60277	.51003	.43191	.36604	.31048
35	.70591	.59387	.50003	.42137	.35538	.29998
36	.69892	.58509	.49022	.41109	.34503	.28983
37	.69200	.57644	.48061	.40107	.33498	.28003
38	.68515	.56792	.47119	.39128	.32523	.27056
39	.67837	.55953	.46195	.38174	.31575	.26141
40	.67165	.55126	.45289	.37243	.30656	.25257
41	.66500	.54312	.44401	.36335	.29763	.24403
42	.65842	.53509	.43530	.35448	.28896	.23578
43	.65190	.52718	.42677	.34584	.28054	.22781
44	.64545	.51939	.41840	.33740	.27237	.22010
45	.63905	.51171	.41020	.32917	.26444	.21266
46	.63273	.50415	.40215	.32115	.25674	.20547
47	.62646	.49670	.39427	.31331	.24926	.19852
48	.62026	.48936	.38654	.30567	.24200	.19181
49	.61412	.48213	.37896	.29822	.23495	.18532
50	.60804	.47500	.37153	.29094	.22811	.17905

Table VI. Continued.

n	4%	4½%	5%	5½%	6%	6½%
1	.96154	.95694	.95238	.94787	.94340	.93897
2	.92456	.91573	.90703	.89845	.89000	.88166
3	.88900	.87630	.86384	.85161	.83962	.82985
4	.85480	.83856	.82270	.80722	.79209	.77732
5	.82193	.80245	.78353	.76513	.74726	.72988
6	.79031	.76790	.74622	.72524	.70496	.68533
7	.75992	.73483	.71068	.68744	.66506	.64351
8	.73069	.70319	.67684	.65160	.62741	.60423
9	.70259	.67290	.64461	.61763	.59190	.56735
10	.67556	.64393	.61391	.58543	.55839	.53273
11	.64958	.61620	.58468	.55491	.52679	.50021
12	.62460	.58966	.55684	.52598	.49697	.46968
13	.60057	.56427	.53032	.49856	.46884	.44102
14	.57748	.53997	.50507	.47257	.44230	.41410
15	.55526	.51672	.48102	.44793	.41727	.38883
16	.53391	.49447	.45811	.42458	.39365	.36510
17	.51337	.47318	.43630	.40245	.37136	.34281
18	.49363	.45280	.41552	.38148	.35034	.32189
19	.47464	.43330	.39573	.36159	.33051	.30224
20	.45639	.41464	.37689	.34274	.31180	.28380
21	.43883	.39679	.35894	.32487	.29416	.26648
22	.42196	.37970	.34185	.30794	.27751	.25021
23	.40573	.36335	.32557	.29188	.26180	.23494
24	.39012	.34770	.31007	.27667	.24698	.22060
25	.37512	.33273	.29530	.26224	.23300	.20714
26	.36069	.31840	.28124	.24857	.21981	.19450
27	.34682	.30469	.26785	.23561	.20737	.18263
28	.33348	.29157	.25509	.22333	.19563	.17148
29	.32065	.27902	.24295	.21169	.18456	.16101
30	.30832	.26700	.23138	.20065	.17411	.15119
31	.29646	.25550	.22036	.19019	.16425	.14196
32	.28506	.24450	.20987	.18038	.15496	.13329
33	.27409	.23397	.19987	.17088	.14619	.12516
34	.26355	.22390	.19035	.16197	.13791	.11752
35	.25342	.21425	.18129	.15353	.13011	.11035
36	.24367	.20503	.17266	.14552	.12274	.10361
37	.23430	.19620	.16444	.13794	.11580	.09729
38	.22529	.18775	.15661	.13074	.10924	.09135
39	.21662	.17967	.14915	.12393	.10306	.08578
40	.20829	.17193	.14205	.11747	.09722	.08054
41	.20028	.16453	.13528	.11134	.09172	.07563
42	.19257	.15744	.12884	.10554	.08653	.07101
43	.18517	.15066	.12270	.10004	.08163	.06668
44	.17805	.14417	.11686	.09482	.07701	.06261
45	.17120	.13796	.11130	.08988	.07265	.05879
46	.16461	.13202	.10600	.08519	.06854	.05520
47	.15828	.12634	.10095	.08075	.06466	.05183
48	.15219	.12090	.09614	.07654	.06100	.04867
49	.14634	.11569	.09156	.07255	.05755	.04570
50	.14071	.11071	.08720	.06877	.05429	.04291

Table VII. Height and Weight — Men.
(Without Clothing)

HEIGHT (Without Shoes) FEET	INCHES	15-19 Yrs. Slender Build	15-19 Yrs. Medium Build	15-19 Yrs. Large Build	20-24 Yrs. Slender Build	20-24 Yrs. Medium Build	20-24 Yrs. Large Build	25-29 Yrs. Slender Build	25-29 Yrs. Medium Build	25-29 Yrs. Large Build	30 Yrs. and Over Slender Build	30 Yrs. and Over Medium Build	30 Yrs. and Over Large Build
4	11	92	102	114	101	112	126	105	117	131	109	121	136
5	0	94	104	117	103	114	128	107	119	134	111	123	138
5	1	96	107	120	105	117	131	109	121	136	113	125	140
5	2	99	110	124	108	120	135	112	124	139	115	128	144
5	3	102	113	127	111	123	138	115	128	144	118	131	147
5	4	105	117	131	114	127	143	119	132	148	122	135	152
5	5	109	121	136	118	131	147	123	136	153	125	139	156
5	6	113	125	140	122	135	152	126	140	157	129	143	161
5	7	116	129	145	125	139	156	130	144	162	132	147	165
5	8	120	133	149	129	143	161	133	148	166	136	151	170
5	9	123	137	154	132	147	165	137	152	171	141	156	175
5	10	128	142	159	136	151	170	141	157	176	145	161	181
5	11	132	147	165	141	156	175	146	162	182	150	167	188
6	0	137	152	171	145	161	181	151	168	189	156	173	194
6	1	141	157	176	150	166	186	157	174	195	161	179	201
6	2	146	162	182	154	171	192	161	179	201	167	185	208

Table VII. Height and Weight — Women.
(Without Clothing)

HEIGHT (Without Shoes) FEET	INCHES	15-19 Yrs.			20-24 Yrs.			25-29 Yrs.			30 Yrs. and Over		
		Slender Build	Medium Build	Large Build	Slender Build	Medium Build	Large Build	Slender Build	Medium Build	Large Build	Slender Build	Medium Build	Large Build
4	8	90	100	113	95	105	117	97	108	122	100	111	125
4	9	91	101	114	96	107	119	99	110	124	102	113	127
4	10	92	102	115	98	109	123	101	112	126	104	115	129
4	11	94	104	117	100	111	125	103	114	128	105	117	132
5	0	96	107	120	103	114	128	104	116	131	107	119	134
5	1	99	110	122	105	117	132	107	119	134	110	122	137
5	2	102	113	127	108	120	135	111	123	138	113	125	141
5	3	104	116	131	111	123	138	113	126	142	116	129	145
5	4	108	120	135	113	126	142	116	129	145	119	132	149
5	5	112	124	140	117	130	146	120	133	149	123	136	153
5	6	115	128	144	121	134	151	123	137	154	126	140	158
5	7	119	132	149	124	138	155	127	141	158	130	144	162
5	8	122	136	153	127	141	159	131	145	163	133	148	167
5	9	126	140	158	131	145	163	134	149	167	136	151	170
5	10	131	145	163	134	149	168	137	152	171	140	155	174
5	11	135	150	168	139	154	173	140	156	176	143	159	179

Index